Dynamics of Large Herbivore Populations
in Changing Environments

T0260475

Dynamics of Large Herbivore Populations in Changing Environments: Towards Appropriate Models

Edited by

Norman Owen-Smith

WILEY-BLACKWELL

A John Wiley & Sons, Ltd., Publication

This edition first published 2010, © 2010 by Blackwell Publishing Ltd

Blackwell Publishing was acquired by John Wiley & Sons in February 2007. Blackwell's publishing program has been merged with Wiley's global Scientific, Technical and Medical business to form Wiley-Blackwell.

Registered office: John Wiley & Sons Ltd, The Atrium, Southern Gate, Chichester, West Sussex, PO19 8SQ, UK

Editorial offices: 9600 Garsington Road, Oxford, OX4 2DQ, UK

The Atrium, Southern Gate, Chichester, West Sussex, PO19 8SQ, UK
111 River Street, Hoboken, NJ 07030-5774, USA

For details of our global editorial offices, for customer services and for information about how to apply for permission to reuse the copyright material in this book please see our website at www.wiley.com/wiley-blackwell

Library of Congress Cataloging-in-Publication Data

Dynamics of large herbivore populations in changing environments: towards appropriate models / edited by Norman Owen-Smith.
    p. cm.
  "This book originated from a working group established at the National Center for Ecological Analysis and Synthesis (NCEAS) at the University of California, Santa Barbara" – Pref.
   ISBN 978-1-4051-9894-3 (hardback) – ISBN 978-1-4051-9895-0 (pbk.)
1. Ungulates – Ecology. 2. Herbivores – Ecology. 3. Mammal populations – Mathematical models. I. Owen-Smith, Norman. II. National Center for Ecological Analysis and Synthesis
  QL737.U4D96 2010
  599.6'17 – dc22

                                        2009038755

ISBN: 9781405198943 (hardback) and 9781405198950 (paperback)

A catalogue record for this book is available from the British Library.

Set in 10.5/12.5 PhotonMT by Laserwords Pvt Ltd, Chennai, India
Printed and bound in Malaysia by Vivar Printing Sdn Bhd

1   2010

Contents

Contributors

Tim Coulson, Department of Life Sciences, Imperial College London, Ascot, UK. Email: T.coulson@imperial.ac.uk.

Marco Festa-Bianchet, Department of Biology, University of Sherbrooke, Sherbrooke, Canada. Email: marco.festa-bianchet@usherbrooke.ca.

Jean-Michel Gaillard, Laboratoire de Biometrie et Biologie Evolutive, Universite Claude Bernard Lyon, Villeurbanne Cedex, France. Email: gaillard@biomserv.univ-lyon1.fr.

Iain J. Gordon, CSIRO Davies Laboratory, Aitkenvale, Queensland, Australia. Email: Iain.Gordon@CSIRO.au.

John Gross, Inventory and Monitoring Program, National Park Service, Fort Collins, Colorado, USA. Email: John_Gross@nps.gov.

Jason Marshal, School of Animal, Plant and Evironmental Sciences, University of the Witwatersrand, Johannesburg, South Africa. Email: Jason.Marshal@wits.ac.za.

Norman Owen-Smith, School of Animal, Plant and Environmental Sciences, University of the Witwatersrand, Johannesburg, South Africa. Email: Norman.Owen-Smith@wits.ac.za.

N. Thompson Hobbs, Natural Resource Ecology Laboratory and Graduate Degree Program in Ecology, Colorado State University, Fort Collins, Colorado, USA. Email: Tom.Hobbs@ColoState.edu.

Preface

This book originated from a working group established at the National Center for Ecological Analysis and Synthesis (NCEAS) at the University of California, Santa Barbara. The problem confronted was the contrasting dynamics shown by populations of large mammalian herbivores in recent decades. Some of these populations had declined abruptly in abundance, most notably in certain protected areas in Africa. Others had become so abundant as to threaten plant populations and forest regeneration, especially in parts of Europe and North America. Conventional models emphasizing population regulation around some steady state seemed inadequate to explain these divergent trends. Hence, the specific aims of the group were to challenge the prevalent models using existing data sets, and develop alternative models capable of accommodating the complexity of environmental influences operating at different spatial and temporal scales.

The group met annually over 3 years, 2001–2003. Much shared understanding was gained, and several joint-authored publications appeared in journals. Numerous other papers have subsequently been published relating to the dynamics of large ungulate populations. At the same time, population ecology has been undergoing a phase of theory maturation, evidenced by the appearance of several books (e.g. Lande, Engen and Saether 2003, Turchin 2003, Cuddington and Beisner 2005) as well as numerous papers in ecological journals. A "balance of nature" perspective is no longer the prevalent paradigm (Cooper 2003), and new approaches provide greater recognition of disequilibrium processes, nonlinear responses, and the consequences of spatial structure, enabled by expanded computational capacity to accommodate this complexity (Owen-Smith 2002, Ellner and Guckenheimer 2006, Clark 2007).

Studies on large herbivores have been making a substantial contribution to this theoretical reassessment. New insights have been provided by the opportunities to follow changes in certain ungulate populations in demographic or even individual detail, revealing response mechanisms that generally remain cryptic for small mammal or insect populations. Being directly dependent on vegetation as a food resource, herbivore populations respond sensitively to patterns of plant growth, and hence to changing climatic and human influences on vegetation dynamics. Complications arise from the additional effects of predation and hunting on some of these populations, as well as from fragmentation of the spatial context within which they exist.

It is time to take stock of the new knowledge that has been gained from these large herbivore studies. In particular, there is a need to assess what revision of theory, expressed through models of population dynamics, is needed to represent and explain the causes of change, looking beyond the expected density-dependent feedbacks that have apparently been ineffective in some circumstances.

Accordingly, this book is aimed at filling this need through (i) reconciling theoretical models with empirical findings on the population dynamics of large mammalian herbivores, and (ii) developing appropriate models for identifying the factors and processes causing changes in abundance. It is structured as follows.

Chapter 1 summarizes findings from a set of long-term studies on herbivore populations that have been especially influential in revealing the processes contributing to changes in abundance. Chapter 2 outlines the suite of modeling approaches representing prevalent theory and concepts. The following chapters then explore particular aspects more comprehensively. Chapter 3 is concerned with identifying local and broad-scale climatic influences on population dynamics, as well as the modifying effects of predation and hunting, structured as a temperate–tropical comparison. Chapter 4 considers the demographic processes that generate changes in abundance, as illuminated by individual-based studies. Chapter 5 identifies circumstances that may lead to irruptive oscillations in abundance. Chapter 6 examines how spatial heterogeneity modifies the influences of temporal variation in conditions on population processes. Finally, Chapter 7 outlines an alternative paradigm for accommodating the effects of spatial and temporal variability in resources and conditions on population dynamics.

All chapters were subject to outside review before being accepted for publication, initially individually and then by the two reviewers of the complete book. Chapters were revised and re-submitted in the light of critical comments. We are indebted particularly to John Fryxell for providing very incisive suggestions for final improvement of all of the chapters. We are also indebted to the various people, both members of the NCEAS group and others from outside this group, who gave some of their time to assess particular chapters. Most fundamentally, the National Center for Ecological Analysis and Synthesis, University of California, Santa Barbara, provided the funding that enabled the working group "Dynamics of large mammalian herbivores in changing environments: alternative modeling approaches" to meet and share our experiences and expertise.

Norman Owen-Smith

1

Definitive case studies

Norman Owen-Smith[1] and Jason P. Marshal[1]

[1]*School of Animal, Plant and Environmental Sciences, University of the Witwatersrand, Johannesburg, South Africa*

Certain long-term studies on large herbivore populations have made especially influential contributions to current understanding of population dynamics. They are outstanding either for the detailed understanding that they have provided based on individually recognizable animals, or for the windows opened into particular processes from the prolonged study duration, or both. This chapter provides a summary outline of the findings that have emerged from these studies, and thus of the particular ways in which studies of large herbivores have advanced our understanding of population processes. It provides the empirical context for the reassessment of theoretical models that is the theme running through the book.

The eight studies assembled all extended over one or more decades. Two have been especially eminent for the wealth of publications produced, including two books: the study of red deer on the Isle of Rum, and of Soay sheep on the Isle of Hirta, both situated off the west coast of Scotland. Two further investigations based likewise on individually identifiable animals have recently become prominent in the literature, focused on roe deer in two regions of France, and bighorn sheep in two localities in Canada. Among African ungulates, a study on greater kudu employing individually recognizable animals was conducted in two regions of South Africa's Kruger National Park over a decade, and extended through a broader spatial and temporal context by park-wide censuses. Another exceptionally long-term

---

Scientific names of species referred to in the text are given in the index.

Dynamics of Large Herbivore Populations in Changing Environments, 1st edition. Edited by Norman Owen-Smith.
© 2010 Blackwell Publishing

study used repeated aerial censuses supported by ground surveys to follow the long-term dynamics of the migratory wildebeest population within the Serengeti ecosystem in Tanzania. Two further studies in North America employing aerial surveys within an ecosystem context documented the long-term dynamics of moose in Isle Royale National Park, and elk in Yellowstone National Park.

Some of the theoretical issues to be borne in mind when evaluating these studies, anticipating the models to be assessed in later chapters, are listed below:

1 How stable, or unstable, have the dynamics of these populations been, as indicated by temporal variation in abundance?
2 How have island restrictions on movements affected population dynamics?
3 How has density dependence been manifested?
4 What climatic influences on abundance have been apparent?
5 What distinctions in demographic responses to these factors are evident?
6 How has predation, or hunting, modified population dynamics?
7 Has vegetation degradation been apparent at high herbivore density?

1.1 Red deer on Rum

This detailed study of red deer (Fig. 1.1) commenced in 1971 in a 12 km² section of the North Block of the Isle of Rum (Clutton-Brock et al. 1982). Culling was suspended within the study area the following year, but continued on the remainder of the island. Although the study population was open to movements, females and their offspring generally remained resident, while males frequently dispersed beyond the study area (Coulson et al. 2004). Predators were absent, apart from golden eagles, and there was little grazing competition from the cattle and goats found elsewhere on the island. Individual deer were identified from natural markings or body shape features, plus collars or ear tags placed on a large proportion of calves soon after birth. Many adult animals were also marked with collars to reduce the likelihood that they would be shot when they moved out of the study area. Censuses of the deer population within the study area were carried out several times each month on foot, facilitated by the prevalently open heathland vegetation.

The number of animals over 1 year in age in the study population grew from about 160 initially to fluctuate between 230 and 370 animals after 1986 (Coulson et al. 2004, Fig. 1.2). The population increase was greater among females than among males, and correspondingly the adult sex ratio

Figure 1.1 Red deer hind on Rum, Scotland (photo: N. Owen-Smith).

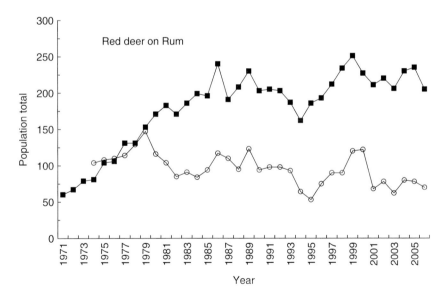

Figure 1.2 Changes in adult red deer population in the North Block of Rum (females: *filled squares*; males: *open circles*).

shifted from 0.5 females per male initially to 1.7 later (Clutton-Brock et al. 1997). The widening skew in the sex ratio was due to greater emigration by locally born males plus reduced immigration by new males. At peak abundance, the deer density in the North Block was 30 animals over 1 year in age per km^2.

The 2.5-fold increase in abundance of the female segment was associated with a decline in fecundity among both young and old females: the proportion of 3-year-old females giving birth dropped from 80% to 5–30%, whilst the annual proportion of females over 9 years in age producing offspring decreased from 95% to 65–80% (Clutton-Brock et al. 1982, Coulson et al. 2004). Calf mortality during summer showed little change, whereas calf mortality during winter rose from initially under 5% (Guinness et al. 1978, Clutton-Brock et al. 1985) to around 40% (annual range 10–80%) (Albon et al. 2000, Coulson et al. 2004). As a result, the ratio of juveniles to adult females the following spring decreased from 48 to 28% (Clutton-Brock et al. 1985). Overwinter mortality among adult females doubled, from 4% per year during the increase phase prior to 1980 to 8.5% per year over the subsequent period of stabilization (Albon et al. 2000). Two severe winters associated with mortality greater than 20% contributed to this elevation in mortality. Mortality among adult males differed similarly between these periods (Kruuk et al. 1999).

Weather conditions affected juvenile mass at birth, with warmer springs promoting heavier calves (Albon et al. 1987). Individuals that weighed more at birth showed higher reproductive success as adults. The proportion of 3-year-old females giving birth was negatively related to precipitation during the preceding late summer, as well as being influenced by winter temperature conditions and the number of days with snow cover (Albon and Clutton-Brock 1988, Langvatn et al. 1996). Conditions in the year of birth also affected the subsequent survival and reproductive success of males (Rose et al. 1998).

Annual variance in the population growth rate was low during the early increase phase, affected mainly by the pregnancy rate among adults (Albon et al. 2000, Coulson and Hudson 2003, Coulson et al. 2004). Only about 80% of females older than 3 years produced offspring each year during this period (Clutton-Brock et al. 1982). After peak abundance had been attained, the annual growth rate fluctuated more widely, with annual variation in adult survival becoming the main contributor, reinforced by strong covariation in overwinter survival of juveniles. This was partly a result of the increased proportion of individuals older than 8 years in the adult segment. Yearling survival also varied widely between years. Conditions experienced in the year of birth affected birth mass, early growth, subsequent survival and reproductive performance as adults (Albon et al. 1987).

Male survival was more variable than that of females at all stages, from birth through adulthood.

The reduction in birth mass associated with rising density indicated food limitation. However, grazing and browsing impacts on the vegetation were not apparent. Females with young concentrated their grazing especially on patches of *Agrostis/Festuca* grassland offering highest quality forage while sufficient amounts remained, and spread into the extensive *Calluna* heathland and patches of *Molinia* grassland during winter (Clutton-Brock et al. 1982). Contesting for food was evident during summer when grazing was locally concentrated, but access to food took the form of a scramble in winter when the deer were more widely dispersed (Coulson et al. 1997). Lags related to instability in the age structure of the population persisted for almost two decades through the period of population stabilization, somewhat beyond the generation time of 8 years. A two-fold difference in contributions by the most and least successful cohorts to population dynamics was evident (Coulson et al. 2004).

This red deer population reached its asymptotic density without much overshoot or vegetation degradation. The oceanic climate with high precipitation but relatively light snow cover helped maintain food availability through winter.

1.2   Soay sheep on Hirta

Soay sheep (Fig. 1.3) are feral survivors of primitive domesticated sheep introduced into the St. Kilda archipelago 2000–3000 years ago. The Hirta

Figure 1.3 Soay sheep ewe on Hirta (photo: T. Coulson).

population was established in the 1930s when humans moved out and sheep were moved across from the nearby Island of Soay. The first studies began in 1955 (Jewell et al. 1974), while more detailed investigations were initiated in 1985, concentrated in the Village Bay region (2.3 km^2) of the 6.4-km^2 island (Clutton-Brock and Pemberton 2004). Annual censuses covering the whole island have been conducted each summer, supported by more frequent counts within the Village Bay area. In the latter area, over 95% of lambs born were caught, weighed, and marked with ear tags during the first month after birth, adding to tags placed earlier, so that the precise age of most animals was known (Clutton-Brock et al. 1991). The total Soay sheep population on Hirta has fluctuated between lows of around 600 and peaks sometimes exceeding 2000 individuals, with the oscillation period typically 3–4 years (Fig. 1.4; Clutton-Brock et al. 1991, Grenfell et al. 1992). The effective population density is extremely high (100–330 sheep per km^2), facilitated by high rainfall supporting lush meadows on volcanic soils fertilized by sheep manure, and enriched in sodium by sea spray (Jewell et al. 1974).

The repetitive oscillations in abundance shown by these sheep have been related to the early timing of births relative to the seasonal cycle in food abundance, plus the early age at first parturition (Clutton-Brock et al. 1991, 1992, 1997, Clutton-Brock and Coulson 2002). Lambs are born

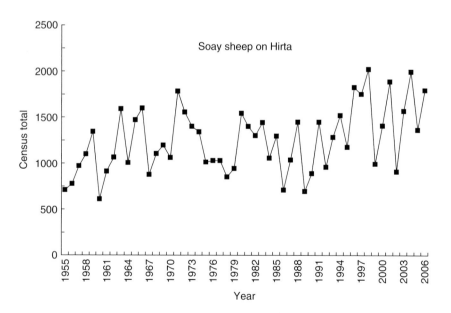

Figure 1.4 Changes in the Soay sheep population on Hirta.

in March preceding the spring growth of grasses, and weaned early in summer, enabling mothers to recover before the food restrictions of winter. First conception generally occurs at 7–8 months, so that most females produce their first offspring by 1 year of age. About 15% of adults produce twin offspring. This high reproductive potential has enabled the population to increase numerically by as much as 50% during the course of summer. The rapid rise in abundance led to depletion of most food by the following winter, with grass biomass dropping to as low as $5–10\,g/m^2$ (Milner and Gwynne 1974). Severe mortality followed when adverse weather conditions occurred towards the end of winter.

Wet and windy weather through February–March, associated with high values of the North Atlantic oscillation (NAO) index, amplified mortality to varying degrees among different population segments after high abundance had been attained (Clutton-Brock et al. 1991, Milner et al. 1999, Catchpole et al. 2000). Population density rather than weather was the overriding influence on the survival of lambs and adult males. Juvenile recruitment at 6 months of age declined from up to 0.45 per adult female at low density to about 0.25 at high density (Clutton-Brock et al. 2004). Survival among adult females was affected by adverse weather but not by density. The proportion of yearling females producing lambs was lower when the autumn density was high. About 80% of adult females gave birth each year irrespective of the population size, but the twinning rate declined with increasing density. The decline in body mass of the sheep during winter was greater in crash years than in other years, indicating food limitation, although nematode parasite loads may have contributed to the mass loss. Wet, stormy weather during the preceding winter as well as high density resulted in a reduction of the birth mass of lambs, and hence reduced survival. In some years, more than 80% of lambs, yearlings, and adult males, more than 50% of yearling females, and up to 45% of adult females of all ages, died during a single winter. Because of this differential mortality, the overall magnitude of the population crash depended on the prior population structure (Catchpole et al. 2000, Coulson et al. 2001). Adverse weather had little impact on mortality in years when the population density remained low following a crash.

Following population crashes, the survivors consisted mainly of prime-aged females. Fecundity remained low in the year after a population crash. Annual mortality among adult females averaged 7.5%, including 45% mortality in 1 crash year (Milner et al. 1999). Cohorts of sheep born after adverse winters, or following winters of high population density, were not only lighter at birth, but reached reproductive maturity at an older age, gave birth at a later date, and were less likely to produce twins than sheep born in other years (Forchhammer et al. 2001).

The synchronizing effect of regional weather was evident from the spatially correlated dynamics of Soay sheep populations on neighboring islands in the St. Kilda group (Grenfell et al. 1998). Long-term effects of the severe grazing pressure on the capacity of the vegetation to support the sheep were not evident. Milder winters may have contributed to the elevated peak densities attained in recent years (Berryman and Lima 2006).

1.3   Roe deer in France

The roe deer (Fig. 1.5) study took place in two forest areas (Gaillard et al. 1993a, 1997). Trois Fontaines (13.6 km^2) in eastern France represented a continental climate with relatively severe winters, but otherwise good quality habitat. Chizé (26.6 km^2) in western France has an oceanic climate with relatively mild winters, but habitat conditions are less favorable due to infertile soils and frequent droughts. There was no predation on adult roe deer, although foxes killed some fawns. Because fences precluded emigration, roe deer numbers within both areas were controlled by annual removals. The density of roe deer at Trois Fontaines was maintained at 15–18 animals per km^2 until 2001, after which the population was

Figure 1.5 Roe deer female in France (photo: B. Hamann).

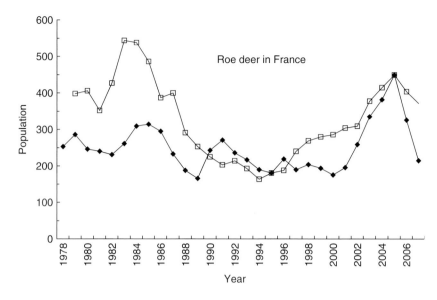

Figure 1.6 Changes in adult roe deer populations in two study areas in France (Trois Fontaine: *filled squares*; Chize: *open squares*).

allowed to grow towards a peak of around 25 animals per km^2 (Fig. 1.6). At Chizé the density increased from 13 to 21 animals per km^2 between 1979 and 1983, then was reduced to around 8 per km^2, but allowed to rise after 2000 towards the earlier density level. To overcome the difficulty in observing the deer in their dense woodland habitat, population dynamics and demography were followed by marking and recapturing animals, commencing in 1976 at Trois Fontaines and in 1978 at Chizé. About half of the marked animals still alive were recaptured annually. From 1985 onward, many neonates were caught and marked within a few days of birth in both sites.

At Trois Fontaines, almost all females aged 2 years or more gave birth each May, producing on average 1.64 fawns per reproductive female (Gaillard et al. 1992, 1997). At Chizé, the proportion of 2-year-old females producing offspring varied among years depending on the population density, and the number of fawn produced per female declined from 1.6 in 1979 to 1.3 in 1985. Early survival through the summer was affected by the population density at Chizé (Gaillard et al. 1997), and at Trois Fontaines after 2001 (Gaillard et al. unpublished data). In both sites, early survival increased with increasing rainfall during spring and early summer (Gaillard et al. 1997). However, heavy rain in April–May during late gestation and cooler summers had a negative influence on the mass of fawns during winter at Chizé (Gaillard et al. 1996). Low fawn survival was associated with reduced birth mass and/or retarded early growth (Gaillard et al. 1993b). However, at Trois Fontaines, fawns born during unfavorable springs were able to compensate for reduced early growth by exploiting the high availability of forage through late summer and early autumn (Gaillard et al. 1993c). Fawns surviving at the end of winter tended to be heavier at Trois Fontaines than at Chizé (Gaillard et al. 1996), and fawn survival was generally lower at Chizé than at Trois Fontaines. Differences in the birth mass and subsequent growth rate gave rise to cohort differences expressed through effects on subsequent survival and reproduction at Chizé (Gaillard et al. 1997).

Adult survival rates were higher for females (93%) than for males (85%) in both study areas (Gaillard et al. 1993a). Adult survival rates decreased beyond 7 years of age more strongly among males than among females, associated with more pronounced tooth wear. Survival between the juvenile and yearling stages (i.e. from 8 to 20 mo of age) was lower than adult survival at Chizé but not at Trois Fontaines, with growth also slower in the former study area. Juvenile survival to 6 months of age averaged 50% at Chizé compared with 63% at Trois Fontaines, with the annual range of variation being 39–91% (Gaillard et al. 1997). Adult survival was reduced in years with severe winters at Trois Fontaines in both sexes. Cohort effects on the body mass attained and consequent reproductive success were long-lasting among animals born under high-density conditions (Pettorelli et al. 2002, Gaillard et al. 2003). Adult survival rates within the prime age range (2–8 yr) appeared insensitive to both changing density and climate, so that variation in juvenile survival through summer was mostly responsible for the annual changes in population abundance, after discounting the effect of culling (Gaillard et al. 1993a, 2000).

1.4 Bighorn sheep in Alberta

The bighorn sheep (Fig. 1.7) study was conducted in two study areas 160 km apart in the eastern ranges of the Rocky Mountains in Alberta, Canada. Observations at Ram Mountain (38 km^2) commenced in 1971, and those at Sheep River (60 km^2) in 1981 (Festa-Bianchet et al. 1997, Jorgenson et al. 1997). The isolated Ram Mountain population was maintained at about 100 animals until 1981 by annual removals. Thereafter it was allowed to grow to a peak of 210 individuals in 1992, following which numbers declined due to density feedbacks plus predation (Fig. 1.8). The study animals at Sheep River represented a segment of a wider metapopulation. Sheep numbers there fluctuated around 150 animals until reduced

Figure 1.7 Bighorn sheep ewe with lamb in Rocky Mountains (photo: F. Pelletier).

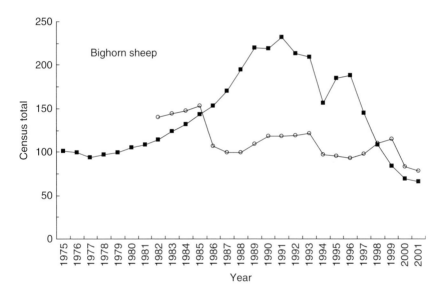

Figure 1.8 Changes in bighorn sheep populations in two study areas in Canada (Ram Mountain: *solid squares*; Sheep River: *open circles*).

by a pneumonia outbreak in 1986. Both study areas were unfenced and open to predation by wolves, cougars, black bears, and coyotes, as well as being subject to trophy hunting of males (Festa-Bianchet et al. 2006). Over 90% of the sheep at Ram Mountain, and up to 94% of those at Sheep River, were marked with ear tags for individual identification. Population surveys took place several times monthly through summer in both study areas until 1987, then four times annually through 1988–93, with survival estimated using capture–recapture methods. At Ram Mountain, sheep were trapped repeatedly in a corral for weighing, while only chest girth measurements on lambs were obtained at Sheep River.

At Ram Mountain, the proportion of lactating 2-year-old females declined shortly after the population began increasing, following the suspension of removals (Festa-Bianchet et al. 1995). A density influence on the overwinter survival of juveniles was apparent in both study areas (Jorgenson et al. 1993, Portier et al. 1998). Mortality among yearling females varied widely between years, and at Ram Mountain was related to population density (Jorgenson et al. 1997), apparently through effects on body mass (Festa-Bianchet et al. 1997). Density effects on adult survival were not detected. Survival rates among prime adult females averaged 94% at Ram Mountain and 92% at Sheep River, in the absence of predation (Festa-Bianchet et al. 2006). Adult survival rates declined beyond 7 years of age more strongly among males than females. Differences in body mass and chest girth at weaning influenced the survival of juvenile sheep (Festa-Bianchet et al. 1997), which was higher on average at Ram Mountain (53%) than at Sheep River (41%). Female longevity was related to their body mass while young, affecting lifetime reproductive success (Bérubé et al. 1999).

At Ram Mountain, warm springs improved the survival of juvenile sheep through summer and over the following winter, but only when population density was high (Portier et al. 1998). Winter temperature also had a positive influence on first-year survival under high-density conditions. High precipitation during spring improved juvenile survival, in this case, independently of density. However, the growth rate of juveniles was lower following springs with a rapid green-up in forage (Pettorelli et al. 2007). Slower snowmelt led to small-scale variability in snow disappearance and hence greater variability in forage green-up and quality among patches. Winter weather did not affect survival beyond the juvenile stage (Jorgenson et al. 1997).

Surges in predation were recorded as a result of the targeting of bighorn sheep by individual cougars causing declines in both sheep populations, at Ram Mountain by two-thirds over 5 years (Festa-Bianchet et al. 2006). The increase in annual mortality among prime-aged females due to predation was from 6 to 10% at Ram Mountain, and from 8 to 20% at Sheep River, with impacts on other demographic segments somewhat greater. Juvenile mortality rose from 47 to 80% at Ram Mountain, and from 59 to 79% at Sheep River.

Figure 1.9 Greater kudu female in Kruger (photo: N. Owen-Smith).

1.5   Kudu in Kruger

The detailed study on greater kudu (Fig. 1.9) was carried out in two areas 70 km apart in the southern part of Kruger National Park, each about 60 km^2 in extent, over a 10-year period (1974–84). Observations covered the growth of these subpopulations towards peak abundance, followed by a drop in numbers associated with a severe drought, then a rebound (Fig. 1.10). All animals in both study areas were individually recognizable from natural markings, documented photographically. Females could be aged to the year of birth from relative body size up to 2 years of age, and males up to 5 years of age from horn shape. The age of an increasing proportion of the study populations became known from photographs of individual animals taken in or shortly after the year of their birth. The annual survival of females plus young was recorded by registering the individuals present in discrete social units during the late dry season, 7–9 months after calves were born. Annual sample sizes were 31–135 in one study area and 62–135 in the other study area, representing about half of the total number of animals on file in these groups within each area. Age-related changes in male survival beyond 2 years of age were estimated from the age structure of the male segment over the study period, corrected for cohort variation and emigration. The maximum density levels attained amounted to almost four animals per km^2 in one study area and around three animals per km^2 in the other. Records of changes in kudu

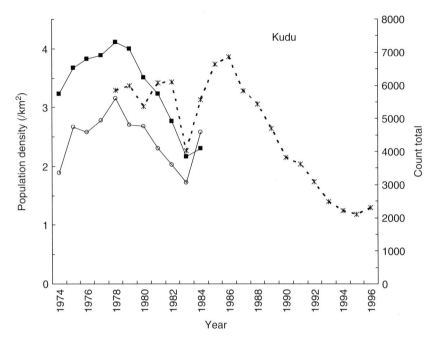

Figure 1.10 Changes in kudu population density in Pretorius Kop (*solid squares*) and Tshokwane (*open circles*) study areas, plus change in kudu population counted in southern half of Kruger National Park (*stars and dotted line*).

abundance were extended by aerial censuses covering almost the entire $19,500\,km^2$ extent of Kruger Park in most years between 1977 and 1996, supported by ground surveys of population structure between 1983 and 1996 (Fig. 1.10).

Annual rainfall variation strongly influenced age-class-specific survival rates (Owen-Smith 1990). Juvenile recruitment represented by the mother : offspring ratio varied annually from between 10 and 80%, with a mean of 45%. Most of the calf losses occurred before the dry season, indicating that rainfall influences on food availability during the wet season affected offspring survival, presumably via the nutritional status of the mother. The prior annual rainfall also influenced survival into the yearling stage (mean 85%) as well as that of females older than 6 years of age (mean 80%). Survival among prime-aged females (2–6 yr) varied little between years, averaging 92% per year. Virtually all mortality was as a result of predation, amounting to 8% per year for females through the prime age range, and 13% for the adult female segment as a whole. Mortality among males was substantially greater than that of females beyond 3 years of age (Owen-Smith 1993). No male survived longer than 10 years, while the oldest female reached 15 years of age. Cold stress evidently reduced the survival of all age classes in a year when exceptionally cold and wet conditions occurred at the end of the dry season (Owen-Smith 2000).

Density effects on survival were revealed only after controlling for rainfall variation (Owen-Smith 1990). While rainfall had the greatest influence on juvenile survival, density effects appeared relatively stronger for older age classes, and were evident even among prime-aged females (Owen-Smith 2006).

Park-wide censuses indicated that kudu numbers in southern Kruger Park rebounded after the 1982–3 drought to reach a peak in 1986, followed by a progressive decline to 20% of the peak abundance several years later (Ogutu and Owen-Smith 2003; Fig. 1.10). Neither annual rainfall nor cold weather explained this trend (Owen-Smith 2000), and there was no density compensation. The declining trend was associated with an elevation in mortality within the adult female segment from 12 to 27% per year, estimated by reconciling the population change with annual juvenile recruitment. Juvenile recruitment estimated from the mother : offspring ratio in ground surveys showed only a minor reduction from 50 to 37% (Owen-Smith and Mason 2005). Enhanced predation was implicated as the most likely cause, both through an increase in the prey base supporting lions (Owen-Smith et al. 2005, Owen-Smith and Mills 2006), and through a shift in prey selection by lions towards alternative prey species including kudu (Owen-Smith and Mills 2008).

1.6   Wildebeest in Serengeti

This study monitored changes in the vast population of migratory wildebeest (Fig. 1.11) moving seasonally between the Serengeti plains in Tanzania and the Mara region of southern Kenya, over a linear distance of 180 km (Boone et al. 2006, Holdo et al. 2009). Starting in 1961, wildebeest aggregations were photographed from the air, and the animals were counted later from the photographs (Campbell and Borner 1995). Recent surveys have been sample-area counts recording other species as well as wildebeest over the full 27,000 km^2 extent of the ecosystem, with sampling intensity increased where wildebeest aggregations were found (Sinclair 1979). Surveys were not undertaken in every year, with a gap of 5 years between 1972 and 1977, and only one count between 1991 and 1999.

Wildebeest numbers increased from an estimated 250,000 in 1961 to reach a peak of about 1.4 million by 1979 (Fig. 1.12). Following a severe drought in 1993, the population decreased to 0.9 million, but

Figure 1.11 Wildebeest aggregation in Serengeti (photo: N. Owen-Smith).

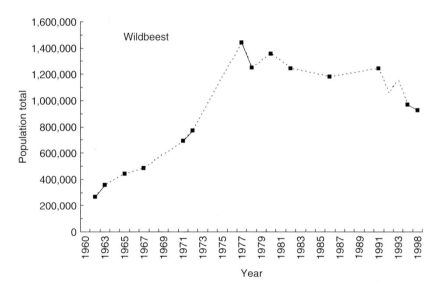

Figure 1.12 Changes in migratory wildebeest population in Serengeti (count totals: *solid squares*; linear interpolation between count totals: *dotted lines*).

thereafter regained its peak abundance, equivalent to a regional density of over 50 animals per km^2. Population structure was assessed from the aerial photographs, distinguishing calves from older animals, plus supporting ground surveys in some years, most comprehensively between 1992 and 1994. Survival rates of juveniles and yearlings were estimated from changes in the ratios of young animals to adult females (>2 yr). Pregnancy rates were estimated from autopsy samples in earlier years, and from radioimmunoassay of fecal samples over 1992–4. Mortality rates of adults were estimated from fresh wildebeest carcasses found during ground surveys, aided by the presence of predators or scavengers on these carcasses as well as the open visibility of the landscape. The number of animals dying each day was expressed as a proportion of the population of live animals in the vicinity.

With rising abundance, the proportion of 2-year-old females producing calves decreased substantially from over 80 to 20%, while pregnancy rates among adult females declined from 95 to 84% (Mduma et al. 1999). Juvenile recruitment averaged 0.27 calves per female at the end of the dry season, which allowing for pregnancy indicates a survival rate of around 40%. Juvenile recruitment fell as low as 0.10 during the 1993 drought, but showed no significant trend with density alone (Owen-Smith 2006). Estimated annual survival among adults averaged 0.91, with a low of 0.75 associated with the severe 1993 drought. Adult mortality rose with increasing density beyond a threshold level. The proportion of adult deaths constituted by older animals increased between 1970 and 1990, but the majority of the dying adult wildebeest remained prime-aged animals. However, the proportion of animals dying that were in poor condition, as revealed by low fat content in bone marrow, increased with density. Wildebeest killed by lions tended to be in better condition, as revealed by bone marrow fat, than those found dead of other causes (Sinclair and Arcese 1995).

The continued population growth of the wildebeest during the 1970s was associated with an increase in amount of rainfall received during the normally dry season months (Pascual and Hilborn 1995). While calf recruitment appeared related to the amount of rainfall received during the dry season months relative to the population density (Mduma et al. 1999), the statistical significance came largely from the low calf survival recorded in the severe drought year (Owen-Smith 2006).

Migration has enabled the wildebeest population to grow to a level where food rather than predation limits its abundance (Fryxell et al. 1988). Illegal harvests by humans, occurring when the animals move outside park boundaries during their migratory circuit, have amounted to around 40,000–50,000 wildebeest annually, equivalent to 3–5% of the population (Mduma et al. 1998). The fairly abrupt increase in adult mortality towards high density may reflect the rapid rise in this source of mortality during the 1980s when anti-poaching efforts were lax. The recent recovery by the population from the post-drought low occurred after better control over the illegal harvests had been achieved by park managers.

Despite the extremely high abundance attained by the wildebeest, there has been no apparent reduction in the productive potential of the grassland. A thickening in density of woody plants, which commonly follows as a result of reduced fire intensity caused by high grazing pressure in savanna ecosystems, appears limited in its extent.

1.7   Moose on Isle Royale

This study has been focused on the interactive dynamics of wolves and moose (Fig. 1.13) in Isle Royale, located in Lake Superior near the United States–Canada border. The $544 \, km^2$ island is managed as a national park with minimal intervention. Observations began in 1958, after a wolf population had become established following earlier colonization of the island by moose (Mech 1966). Moose population estimates for 1959–93 were derived from recoveries of dead moose, while subsequent estimates were obtained from aerial counts conducted during winter when moose were readily visible against the snow. The moose population grew towards a projected peak of around 1500 animals by 1972, then declined in association with a growing wolf population (Fig. 1.14). Following a crash in wolf numbers from 50 to 14 animals by 1981 caused by viral infection, the moose population increased to a high of almost 2500 animals by 1995.

Figure 1.13 Moose female on Isle Royale (photo: J. Vucetich).

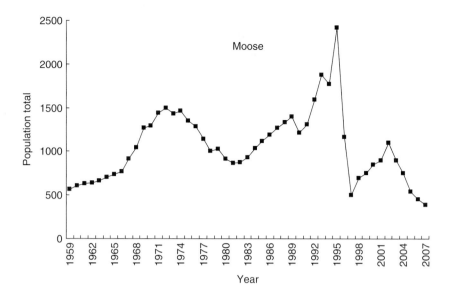

Figure 1.14 Changes in moose population in Isle Royale National Park.

Severe mortality over two successive winters then reduced the moose population by 80% (Peterson 1995, Smith et al. 2003). Recovery from this low point has since been suppressed by a combination of hot summers, abundant ticks, and elevated wolf predation, associated with a persistently low calf proportion.

Early observations suggested that top-down control through the lagged numerical response of the wolves had restricted the irruptive potential of the moose population (Mech 1966). Wolf abundance responded largely to changes in the number of moose over 9 years of age, together with the availability of moose calves, since success in killing prime-aged moose in good condition was low. Snow depth additionally affected the susceptibility of moose to predation, and had persistent effects on twinning, birth mass, and calf survival (Mech et al. 1987, Messier 1991). Calf : adult female ratios in summer varied between extremes of 18 and 58%. Snow conditions were in turn a consequence of broadscale weather patterns associated with the NAO (Post and Stenseth 1998). Severe winters had a negative effect on population growth in the same year, but a positive response the subsequent year due to the benefits of improved food availability in summer on reproductive success. Reductions in survival rates among juvenile or adult moose in response to increasing density were either lacking or insufficient to dampen the sustained growth of the moose population when wolf numbers were low (Peterson 1999). The eventual crash in moose abundance through 1995–6 was associated with extreme weather conditions, exacerbated by an outbreak of ticks (DelGiudice et al. 1997).

Short-term oscillations in moose abundance appear to be related mainly to the availability of winter food (Vucetich and Peterson 2004), with the predator–prey interaction being influential over a longer period. The growth of the moose population has had a substantial negative impact on the abundance of balsam fir (*Abies balsamea*), a major winter food source for the moose (McLaren and Peterson 1994, Post et al. 1999). The proportional representation of balsam fir in the forest canopy declined from 46% around 1850, before the moose arrived, to under 5% (McLaren and Peterson 1994). Nevertheless, abundant regeneration of these firs is evident in the eastern section of the island.

1.8   Elk in North Yellowstone

In the late 1960s, the management policy of the US National Parks Service for Yellowstone National Park shifted towards allowing natural regulatory mechanisms to operate. The largest elk (Fig. 1.15) subpopulation in the north of the park had previously been restricted to around 6000 animals by culling (Houston 1982, Coughenour and Singer 1996, Wagner 2006, Eberhardt et al. 2007). These elk continued to be exposed to hunting during winter when they moved beyond the boundaries of the park, with the offtake increasing to around 25% of the female segment outside the

Figure 1.15   Elk hind in Rocky Mountain National Park (photo: D. Lutsey).

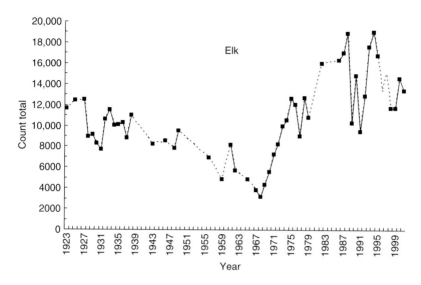

Figure 1.16 Changes in elk population in North Yellowstone (count totals: *solid squares*; interpolation between counts: *dotted lines*).

park between 1976 and 2004. Aerial counts of the winter range accounted for about 75% of the total population. Population classifications made from the helicopter or the ground distinguished calves and yearlings from older animals. The elk reached a peak abundance of 23,000 in 1988, but declined progressively after the reintroduction of wolves into the park in 1995 to 12,000 by 2004 (Fig. 1.16). The summer range occupied by the northern elk subpopulation exceeds 5000 km^2, so that the peak population density amounted to a little over four animals per km^2. During winter, these animals concentrate within a reduced range of 1520 km^2 at an effective density of up to 15 animals per km^2, thereby greatly increasing the browsing pressure in these bottomlands.

Density-related feedbacks on population growth became apparent as the elk population increased towards its peak abundance (Houston 1982). Pregnancy rate among adult females dropped from 87 to 61% (Coughenour and Singer 1996, Singer et al. 1997), and calf survival decreased both through summer and over the following winter, associated with a lowered birth mass. Predation accounted for over 40% of the calf losses. A density feedback on the survival of the adult segment became apparent only towards higher density levels.

Precipitation in the form of rain, especially during spring, positively influenced calf survival during summer as well as survival among both juveniles and adults through the following winter (Coughenour and Singer 1996), and consequently also overall population growth (Taper and Gogan 2002). Juvenile recruitment, indexed by the calf : cow ratio after the birth pulse, depended on precipitation 2 years earlier. Severe winter weather was associated with elevated mortality in late winter, particularly when ice crusts formed on deep snow (Singer et al. 1997, Smith et al. 2003, Evans et al. 2006). Exceptionally severe conditions experienced in the winter of 1996 resulted in the death of most calves and reduced survival among females older than 10 years (Garrott et al. 2003).

Predation losses to coyotes, grizzly and black bears were concentrated especially on light and late-born juveniles (Singer et al. 1997). Wolves targeted mainly young animals and old females (Evans et al. 2006, Wright et al. 2006). The mean age of the adult female elk killed by wolves was almost 14 years, compared with 6.5 years for those shot by human hunters. Models incorporating these patterns suggested that wolf predation was largely compensatory, and that human harvests coupled with adverse climatic conditions were mostly responsible for the elk decline after 1995 (Vucetich et al. 2005). However, indirect consequences of the risk of predation on habitat use and activity could have effects on the elk population not accounted for in this model (Creel et al. 2007, Kauffmann et al. 2007). The annual survival rate of adult female elk fell from 92% prior to 1994, when mortality was largely due to hunting, to 82% after 2000, following the introduction of wolves (Coughenour and Singer 1996, White and Garrott 2005, Evans et al. 2006). Juvenile recruitment estimated by the number of calves per adult female towards the end of winter declined from a mean of 0.27 prior to 1994 to 0.17 after 2000, despite the reduction in population density (Barber-Meyer et al. 2008). Survival rates would be about 10% higher, allowing for pregnancy rates which remained unchanged. However, extremely dry conditions in recent years probably contributed to the elevated mortality rates.

Vegetation in northern Yellowstone has been drastically altered by the browsing effects of elk (Wagner 2006). Aspen, sagebrush, and willow, which were abundant at the turn of the century and contributed substantially to the winter forage of the elk, have been greatly depressed through heavy browsing (Romme et al. 1995, Kay 1997, Ripple and Larsen 2000, Zeigenfuss et al. 2002). Consequences for the capacity of the vegetation to support the elk population have not been assessed.

1.9   Overview

The set of eight studies encompass a wide range of ungulate species, feeding types, and body sizes, but is weighted strongly towards north temperate environments (Table 1.1). Study areas varied vastly in their extent, with those based on individual recognition tending to be in restricted areas and those with the widest spatial coverage tending to show the greatest temporal duration. The two most intensive studies, on red deer and Soay sheep respectively, were conducted in island situations precluding dispersal, although in the case of red deer, animals were able to move out of the study area into other parts of the island. Predation was also lacking in these islands, although management culling restricted red deer numbers outside the study area. Maximum population biomass levels varied over an order of magnitude, from under $5\,kg\,ha^{-1}$ to over $50\,kg\,ha^{-1}$. All studies spanned at least twofold variation in population density levels.

Two of the populations studied have manifested repeated oscillations in abundance, with the attainment of peak numbers being followed by severe mortality, generally in association with adverse weather conditions. The period between successive peaks has been 3–4 years for Soay sheep, but 25 years or more for moose. Both populations exist on islands, but of vastly differing area. The population crashes by moose have been associated with progressive reduction in the browse resource upon which they depend, while the extreme peak biomass levels attained by the Soay sheep have not led to any apparent reduction in the productive potential of their grass forage. In contrast, the red deer population studied, also within an island context, has stabilized in abundance with only minor fluctuations around the asymptotic density. Substantial reductions in pregnancy rates, as well as in both juvenile and adult survival, have contributed towards suppressing the population growth of these deer. The Soay sheep intrinsically have a much higher population growth potential than red deer. Females can produce their first offspring at 1 year of age, while 20% of adults give birth to twins, enabling the population to increase by 50% during the course of a year under ideal conditions. Accordingly, the population level attained by the end of summer may be much greater than what food resources can support through winter, especially when adverse weather intervenes (Clutton-Brock et al. 1997, Clutton-Brock and Coulson 2002). However, moose show irruptive dynamics on Isle Royale despite much slower growth towards reproductive maturity, although a proportion of females do similarly produce twins. The roe deer populations studied also occupy effective islands without predators because of fence restrictions on dispersal from the forest patches, but management culling has restricted the density levels attained.

The kudu population in Kruger Park has shown wide fluctuations in abundance almost equal to the range shown by the Soay sheep, but in this case the declines were associated with perturbations related to rainfall variation, and more recently to an elevation in predation pressure. Rather than

Table 1.1 Summary information for the set of studies

Species	Body mass[1] (kg)	Study area	Extent (km²)	Duration (yrs)	Density (/km²)	Peak biomass[2] (kg/ha)	Indiv. recog.	Dis-persal	Pre-dation	Hunt/cull	Survival rates		Density feedback			Weather Influence		Veget. degrad.
											AdF	Juv	AdF surviv	Juv surviv	AdF fecund	Precip.	Temp.	
Red deer	65	Rum	12	37+	15–33	15	X	(0)	0	(0)	0.96–0.915 (→0.79)	0.75 (→0.3)	X	X	X	-ve	+ve	0
Soay sheep	24	Hirta	6.4	50/23+	100–330	55	X	0	0	0	0.925 (→0.55)	0.45→ 0.2	0	X	X	-ve	+ve	0
Roe deer	30	France	14+26	32+	8–25	5.5	X	0	0	X	0.93	0.56 (0.4–0.9)	0	X	X	+ve	-ve	0
Bighorn sheep	72	Canada	38+60	38+	2–5	2.8	X	X	X	X	0.93	0.47	0	X	X	+ve	+ve	0
Kudu	180	Kruger Park	60+60	10/22	2–4	5.4	X	X	X	0	0.92/0.87 (→0.73)	0.55 (0.1–0.8)	X	X	X	+ve	+ve	0
Wilde-beest	160	Seren-geti	27,000	47+	10–52	62	0	X	X	(X)	0.91 (→0.75)	0.40 (→0.1)	X	0	X	+ve		0
Moose	400	Isle Royale	544	50+	1–4	12	0	0	X	0						-ve		X
Elk	250	Yellow-stone	5,200	85+	1–4	7.5	0	X	X	X	0.92 (→0.80)	0.30 (→0.19)	X	X	X	+/-	+ve	X

X = presence of factor or effect, 0 = absent or no effect, – = no information, +ve = positive influence, -ve = negative influence, brackets indicate equivocal assignments.
[1] mean adult female mass; [2] assumes mean individual mass = three-quarters of adult female mass, AdF = adult female.

being stable in abundance, this population appears resilient through showing fairly rapid recovery from these disturbances. The elk population in northern Yellowstone has also shown widely variable abundance, but mostly related to recovery from the low level previously maintained by management culling. It is uncertain where the recent downward trend following the introduction of a major predator will lead. Contributing to the uncertainty are indications that the adverse vegetation changes induced by elk browsing are being reversed in the presence of the predator. In contrast, the wildebeest population in Serengeti increased from relatively low density to stabilize with only minor fluctuations around its enormous biomass density, despite a huge abundance and diversity of predators. However, its wide migratory movements have evidently enabled it to escape predator limitation. Adverse vegetation changes have not resulted from the huge grazing pressure, perhaps because of the migratory rotation. Predation also emerged as a disrupting influence on both study populations of bighorn sheep, when a specific predator started targeting this prey species.

Red deer on Rum, and wildebeest in Serengeti increased from relatively low density to stabilize with minor fluctuations around an upper abundance asymptote. The trend shown by elk in Yellowstone resembles this pattern, but it remains unclear whether an upper abundance level is being maintained, or a predator-related decline being initiated. For both kudu in Kruger and bighorn sheep in Alberta, equilibration around a maximum abundance level was disrupted by elevated predation losses. Soay sheep on Hirta have exhibited persistent oscillations in abundance over a three-fold range, with a period between peaks of only 3–4 years. Moose on Isle Royale have increased to high abundance followed by population crashes at intervals of 25 or more years, with the most recent case being the most extreme. For roe deer in France, the suspension of removals has been too recent to reveal the intrinsic population trend.

Density feedbacks counteracting population growth were detected in all populations, except for moose, where there has been no critical assessment. Fecundity declined with increasing density, most commonly through a delay in the age at first parturition, although pregnancy rates among adult females were also lowered for red deer, elk, and wildebeest. Juvenile survival was also negatively influenced by rising density, except for wildebeest where a density influence could not be detected when the direct effects of changing food availability were taken into account. Only some of these populations showed reduced survival among adult females, generally in interaction with extreme weather after high-density levels had been attained. Adult males seemed more sensitive to extreme density or weather conditions than females, in those cases where information was available. Density effects were associated with reductions in birth mass and subsequent growth, indicating malnutrition as the underlying cause. However, the density feedback became apparent at widely differing biomass density levels among these populations (Table 1.1).

Warmer temperatures generally had a positive influence on juvenile survival and hence population growth, except for roe deer where cooler summers led to reduced body mass and hence survival among fawns. Effects of lower winter temperatures on northern ungulates were generally not clearly separated from those of snow cover. Kudus showed reduced survival during unusually cold and wet weather at the end of dry season at temperature levels that would be benign for northern ungulates. Increased precipitation in the form of rain during summer generally had a positive influence on calf survival, although not for red deer on Rum. For wildebeest, rain received during the normally dry season was most influential. Greater snowfall during winter negatively affected moose and elk populations occupying regions where snow depth restricted movements. A combination of extreme weather and density conditions in the year of birth had long-lasting influences on the subsequent fecundity of red deer, Soay sheep, and roe deer females.

Mean mortality rates among adult females varied within a surprisingly narrow range of 6–8% per annum, excluding the early growth phase of the red deer population (Table 1.1). However, comparisons are affected by whether survival was estimated across all adult females, including mortality due to senescence, or only through the prime age range. Mortality among adult females in worst years increased to over 20%, but in the case of Soay sheep to as much as 45%. Juvenile mortality prior to weaning appeared somewhat more variable over a range of 44–70%, increasing to as much as 90% under worst conditions. In this case, comparisons are blurred by the stage at which survival was measured and by whether pregnancy rates were taken into account when estimating survival from mother : offspring ratios. Studies following recognizable individuals recorded reduction in survival rates of females beyond some prime age range, with lowered fecundity also evident among old red deer females.

Rather than controlling population growth via a density-related feedback, changes in predation pressure had a disruptive influence on the dynamics of moose, bighorn sheep, and kudu populations. The ungulate populations exposed to substantial predation did not show lower survival rates than those lacking predation, except for Yellowstone elk where mortality among both adults and calves rose substantially following the introduction of wolves. However, for these elk, an additional influence from adverse weather on survival rates cannot be discounted. Culling conducted by the management authority controlled the size of the roe deer population, and perhaps contributed to the stabilization of the red deer subpopulation in the study area through the removal of animals from elsewhere on the island. Controlled hunting had no obvious effect on the dynamics of bighorn sheep, while hunting (now curtailed) probably contributed to the initial decline of elk in Yellowstone following the introduction of wolves. Illegal killing of wildebeest moving outside park boundaries during their mass migration may have contributed to the stabilization of this population.

The growth of the elk population in Yellowstone has clearly had a massive influence on woody plant populations, but with no obvious consequences for the dynamics of the elk. The increase of moose in Isle Royale has also substantially influenced woody vegetation composition, but with no apparent influence on the dynamics of the moose population. Despite the extremely high biomass levels attained by wildebeest in Serengeti, Soay sheep on Hirta, and red deer on Rum, the productive potential of the grass layer upon which they depend does not seem to have been adversely affected.

The attainment of high, relatively stable, abundance levels by wildebeest in Serengeti and red deer on Rum represents the ideal situation that managers desire, but for red deer there was some assistance from management removals, and for wildebeest via illegal hunting. The pattern shown by Soay sheep on Hirta and moose on Isle Royale is what managers fear, i.e. growth to extremely high density levels curtailed by starvation-related die-offs. This is the justification for the removals by live capture or hunting carried out for roe deer and red deer outside the intensive study area. However, while damage to woody vegetation was a worry for Isle Royale, this was not a concern for the lush meadows grazed by the sheep on their wet and fertile island. The fluctuating abundance shown by kudu in Kruger in response to both weather and predation did not warrant any intervention. How managers should respond to the widely variable abundance of elk in Yellowstone, with both predation and weather involved, is currently highly contentious (White and Garrott 2005, Wagner 2006).

The empirical data documented in this chapter provide the basis for models to project the variable influences of weather, population density, predation, and human interventions on the dynamics of large herbivore populations in changing environments. Several of these studies have been especially revealing of the demographic mechanisms underlying population changes, manifested through the responses of particular population segments to these conditions. Subsequent chapters will assess the extent to which the patterns revealed by these case studies can be generalized. However, the immediate need is to assess the modeling concepts and theory that have been developed for the interpretation of these population dynamics, which is the topic of the next chapter.

Acknowledgments

Our summaries of these case studies were checked for accuracy by the following people: red deer plus Soay sheep – Tim Coulson; roe deer – Jean-Michel Gaillard; bighorn sheep – Marco Festa-Bianchet; kudu – Norman Owen-Smith; wildebeest – John Fryxell; moose – John Vucetich. Atle Mysterud, John Fryxell, and Sophie Grange provided helpful critical comments.

References

Albon, S. D. and T. H. Clutton-Brock. 1988. Climate and the population dynamics of red deer in Scotland. In *Ecological Changes in the Uplands*, eds. M. B. Usher and D. B. A. Thompson, pp. 93–107. Blackwell, Oxford.

Albon, S. D., T. H. Clutton-Brock, and F. E. Guinness. 1987. Early development and population dynamics in red deer. II. Density-independent effects and cohort variation. *Journal of Animal Ecology* 56: 69–81.

Albon, S. D., T. N. Coulson, D. Brown, F. E. Guinness, J. M. Pemberton, and T. H. Clutton-Brock. 2000. Temporal changes in key factors and key age groups influencing the population dynamics of female red deer. *Journal of Animal Ecology* 69: 1099–1110.

Barber-Meyer, S. M., L. D. Mech, and P. J. White. 2008. Elk Calf Survival and Mortality Following Wolf Restoration to Yellowstone National Park. *Wildlife Monographs* No. 169, 30pp.

Berryman, A., and M. Lima. 2006. Deciphering the effects of climate on animal populations: diagnostic analysis provides new interpretation of Soay sheep dynamics. *American Naturalist* 168: 784–795.

Bérubé, C. H., M. Festa-Bianchet, and J. T. Jorgenson. 1999. Individual differences, longevity, and reproductive senescence in bighorn ewes. *Ecology* 80: 2555–2565.

Boone, R. B., S. J. Thirgood, and J. G. Hopcraft. 2006. Serengeti wildebeest migratory patterns modeled from rainfall and new vegetation growth. *Ecology* 87: 1987–1994.

Campbell, K. and M. Borner. 1995. Population trends and distribution of Serengeti herbivores: implications for management. In *Serengeti II*, ed. A. R. E. Sinclair and P. Arcese, pp. 117–145. University of Chicago Press, Chicago.

Catchpole, E. A., B. J. T. Morgan, T. N. Coulson, S. N. Freeman, and S. D. Albon. 2000. Factors influencing Soay sheep survival. *Applied Statistics* 49: 453–472.

Clutton-Brock, T. H., and T. Coulson. 2002. Comparative ungulate dynamics: the devil is in the details. *Philosophical Transactions of the Royal Society B: Biological Sciences* 357: 1285–1298.

Clutton-Brock, T. H., B. T. Grenfell, and T. Coulson. 2004. Population dynamics in Soay sheep. In *Soay Sheep. Dynamics and Selection in an Island Population*, eds. T. Clutton-Brock and J. Pemberton, pp. 52–88. Cambridge University Press, Cambridge.

Clutton-Brock, T. H., F. E. Guinness, and S. D. Albon. 1982. *Red Deer: Behavioural Ecology of Two Sexes*. University of Chicago Press, Chicago.

Clutton-Brock, T. H., M. Major, and F. E. Guinness. 1985. Population regulation in male and female red deer. *Journal of Animal Ecology* 54: 831–846.

Clutton-Brock, T. H., O. F. Price, S. D. Albon, and P. A. Jewell. 1991. Persistent instability and population regulation in Soay sheep. *Journal of Animal Ecology* 60: 593–608.

Clutton-Brock, T. H. and J. M. Pemberton (eds). 2004. *Soay Sheep: Dynamics and Selection in an Island Population*. Cambridge University Press, Cambridge.

Clutton-Brock, T. H., O. F. Price, S. D. Albon, and P. A. Jewell. 1992. Early development and population fluctuations in Soay sheep. *Journal of Animal Ecology* 61: 381–396.

Clutton-Brock, T. H., K. E. Rose, and F. E. Guinness. 1997. Density-related sexual selection in red deer. *Proceeding of the Royal Society B: Biological Sciences* 264: 1509–1516.

Coughenour, M. B. and F. J. Singer. 1996. Elk population processes in Yellowstone National Park under the policy of natural regulation. *Ecological Applications* 6: 573–593.

Coulson, T. N., S. Albon, F. Guinness, J. Pemberton, and T. Clutton-Brock. 1997. Population substructure, local density, and calf winter survival in red deer. *Ecology* 78: 852–863.

Coulson, T., E. A. Catchpole, S. D. Albon, et al. 2001. Age, sex, density, winter weather, and population crashes in Soay sheep. *Science* 292: 1528–1531.

Coulson, T., F. Guinness, J. Pemberton, and T. Clutton-Brock. 2004. The demographic consequences of releasing a population of red deer from culling. *Ecology* 85: 411–422.

Coulson, T. and E. Hudson. 2003. When is the birth rate the key factor associated with population dynamics? In *Reproductive Science and Integrated Conservation*, eds. W. Holt, A Pickford J. Rodger, and D. Wildt, pp. 114–130. Cambridge University Press, Cambridge.

Creel, S., D. Christianson, S. Liley, and J. A. Winnie. 2007. Predation risk affects reproductive physiology and demography of elk. *Science* 315: 960.

DelGiudice, G. D., R. O. Peterson, and W. M. Samuel. 1997. Trends of winter nutritional restriction, ticks, and numbers of moose on Isle Royale. *Journal of Wildlife Management* 61: 895–903.

Eberhardt, L. L., P. J. White, R. A. Garott, and D. B. Houston. 2007. A seventy-year history of trends in Yellowstone's northern elk herd. *Journal of Wildlife Management* 71: 594–602.

Evans, S. B., L. D. Mech, P. J. White, and G. A. Sergeant. 2006. Survival of adult female elk in Yellowstone following wolf restoration. *Journal of Wildlife Management* 70: 1372–1378.

Festa-Bianchet, M., T. Coulson, J.-M. Gaillard, J. T. Hogg, and F. Pelletier. 2006. Stochastic predation events and population persistence in bighorn sheep. *Proceedings of the Royal Society B: Biological Sciences* 273: 1537–1543.

Festa-Bianchet, M., J. T. Jorgenson, C. H. Bérubé, C. Portier, and W. D. Wishart. 1997. Body mass and survival of bighorn sheep. *Canadian Journal of Zoology* 75: 1372–1379.

Festa-Bianchet, M., J. T. Jorgenson, M. Lucherini, and W. D. Wishart. 1995. Life-history consequences of variation in age of primiparity in bighorn ewes. *Ecology* 76: 871–881.

Forchhammer, M. C., T. H. Clutton-Brock, J. Lindström, and S. D. Albon. 2001. Climate and population density induce long-term cohort variation in a northern ungulate. *Journal of Animal Ecology* 70: 721–729.

Fryxell, J. M., J. Greever, and A. R. E. Sinclair. 1988. Why are migratory ungulates so abundant? *American Naturalist* 131: 781–798.

Gaillard, J.-M., J. M. Boutin, D. Delorme, G. van Laere, P. Duncan, and J. D. Lebreton. 1997. Early survival in roe deer: causes and consequences of cohort variation in two contrasted populations. *Oecologia* 112: 502–513.

Gaillard, J.-M., D. Delorme, J. M. Boutin, G. van Laere, and B. Boisaubert. 1996. Body mass of roe deer fawns during winter in 2 contrasting populations. *Journal of Wildlife Management* 60: 29–36.

Gaillard, J.-M., D. Delorme, J. M. Boutin, G. van Laere, B. Boisaubert, and R. Pradel. 1993a. Roe deer survival patterns: a comparative analysis of contrasting populations. *Journal of Animal Ecology* 62: 778–791.

Gaillard, J.-M., D. Delorme, and J. M. Jullien. 1993b. Effects of cohort, sex, and birth date on body development of roe deer fawns. *Oecologia* 94: 57–61.

Gaillard, J.-M., D. Delorme, J. M. Jullien, and D. Tatin. 1993c. Timing and synchrony of births in roe deer. *Journal of Mammalogy* 74: 738–744.

Gaillard, J.-M., M. Festa-Bianchet, N. G. Yoccoz, A. Loison, and C. Toïgo. 2000. Temporal variation in fitness components and population dynamics of large herbivores. *Annual Review of Ecology and Systematics* 31: 367–393.

Gaillard, J.-M., A. Loison, C. Toïgo, D. Delorme, and G. van Laere. 2003. Cohort effects and deer population dynamics. *Ecoscience* 10: 412–420.

Gaillard, J.-M., A. J. Sempere, J. M. Boutin, G. van Laere, and B. Boisaubert. 1992. Effects of age and body-weight on the proportion of females breeding in a population of roe deer (*Capreolus capreolus*). *Canadian Journal of Zoology* 70: 1541–1545.

Garrott, R. A., L. L. Eberhardt, P. J. White, and J. Rotella. 2003. Climate-induced variation in vital rates of an unharvested large-herbivore population. *Canadian Journal of Zoology* 81: 33–45.

Grenfell, B. T., O. F. Price, S. D. Albon, and T. H. Clutton-Brock. 1992. Overcompensation and population cycles in an ungulate. *Nature* 355: 823–826.

Grenfell, B. T., K. Wilson, B. F. Finkenstadt, et al. 1998. Noise and determinism in synchronized sheep dynamics. *Nature* 394: 674–677.

Guinness, F. E., T. H. Clutton-Brock, and S. D. Albon. 1978. Factors affecting calf mortality in red deer. *Journal of Animal Ecology* 47: 817–832.

Holdo, R. M., R. D. Holt, and J. M. Fryxell. 2009. Opposing rainfall and plant nutritional gradients best explain the wildebeest migration in the Serengeti. *American Naturalist* 173: 431–445.

Houston, D. B. 1982. *The Northern Yellowstone Elk: Ecology and Management*. MacMillan, New York.

Jewell, P. A., C. Milner, and J. M. Boyd. 1974. *Island Survivors: the Ecology of the Soay Sheep of St. Kilda*. Athlone Press, London.

Jorgenson, J. T., M. Festa-Bianchet, J.-M. Gaillard, and W. D. Wishart. 1997. Effects of age, sex, disease, and density on survival of bighorn sheep. *Ecology* 78: 1019–1032.

Jorgenson, J. T., M. Festa-Bianchet, and W. D. Wishart. 1993. Harvesting bighorn ewes: consequences for population size and trophy ram production. *Journal of Wildlife Management* 57: 429–435.

Kauffmann, M. J., N. Varley, D. W. Smith, D. R. Stahler, D. R. MacNulty, and M. S. Boyce. 2007. Landscape heterogeneity shapes predation in a newly restored predator-prey system. *Ecology Letters*. 10: 690–700.

Kay, C. E. 1997. Viewpoint: ungulate herbivory, willows, and political ecology in Yellowstone. *Journal of Range Management* 50: 139–145.

Kruuk, L. E. B., T. H. Clutton-Brock, S. D. Albon, J. M. Pemberton, and F. E. Guinness. 1999. Population density affects sex ratio variation in red deer. *Nature* 399: 459–461.

Langvatn, R., S. D. Albon, T. Burkey, and T. H. Clutton-Brock. 1996. Climate, plant phenology and variation in age of first reproduction in a temperate herbivore. *Journal of Animal Ecology* 65: 653–670.

McLaren, B. E. and R. O. Peterson. 1994. Wolves, moose, and tree rings on Isle Royale. *Science* 266: 1555–1558.

Mduma, S. A. R., R. Hilborn, and A. R. E. Sinclair. 1998. Limits to exploitation of the Serengeti wildebeest and the implications for its management. In *Dynamics of Tropical Communities*, eds. D. M. Newberry, H. B. Underwood, and H. T. T. Prins, pp. 243–265. Blackwell, Oxford.

Mduma, S. A. R., A. R. E. Sinclair, and R. Hilborn. 1999. Food regulates the Serengeti wildebeest: a 40-year record. *Journal of Animal Ecology* 68: 1101–1122.

Mech, L. D. 1966. *The Wolves of Isle Royale*, National Park Service Fauna Series No. 7. US Government Printing Office, Washington, DC.

Mech, L. D., R. E. McRoberts, R. O. Peterson, and R. E. Page. 1987. Relationship of deer and moose populations to previous winter's snow. *Journal of Animal Ecology* 56: 615–627.

Messier, F. 1991. The significance of limiting and regulating factors on the demography of moose and white-tailed deer. *Journal of Animal Ecology* 60: 377–393.

Milner, J. M., D. A. Elston, and S. D. Albon. 1999. Estimating the contributions of population density and climatic fluctuations to interannual variation in survival of Soay sheep. *Journal of Animal Ecology* 68: 1235–1247.

Milner, C. and D. Gwynne. 1974. The Soay sheep and their food supply. In Island Survivors. *The Ecology of the Soay Sheep of St. Kilda*, eds. P. A. Jewell, C. Milner, and J. Morton-Boyd, pp. 273–325. Athlone Press, University of London.

Ogutu, J. O. and N. Owen-Smith. 2003. ENSO, rainfall and temperature influences on extreme population declines among African savanna ungulates. *Ecology Letters* 6: 412–419.

Owen-Smith, N. 1990. Demography of a large herbivore, the greater kudu *Tragelaphus strepsiceros*, in relation to rainfall. *Journal of Animal Ecology* 59: 893–913.

Owen-Smith, N. 1993. Comparative mortality rates of male and female kudus: the costs of sexual size dimorphism. *Journal of Animal Ecology* 62: 428–440.

Owen-Smith, N. 2000. Modeling the population dynamics of a subtropical ungulate in a variable environment: rain, cold and predators. *Natural Resource Modeling* 13: 57–87.

Owen-Smith, N. 2006. Demographic determination of the shape of density dependence for three African ungulate populations. *Ecological Monographs* 76: 93–109.

Owen-Smith, N. and D. R. Mason. 2005. Comparative changes in adult vs. juvenile survival affecting population trends of African ungulates. *Journal of Animal Ecology* 74: 762–773.

Owen-Smith, N., D. R. Mason, and J. O. Ogutu. 2005. Correlates of survival rates for 10 African ungulate populations: density, rainfall and predation. *Journal of Animal Ecology* 74: 774–788.

Owen-Smith, N. and M. G. L. Mills. 2006. Manifold interactive influences on the population dynamics of a multispecies ungulate assemblage. *Ecological Monographs* 76: 73–92.

Owen-Smith, N. and M. G. L. Mills. 2008. Shifting prey selection generates contrasting herbivore dynamics within a large-mammal predator-prey web. *Ecology* 89: 1120–1133.

Pascual, M. A. and R. Hilborn. 1995. Conservation of harvested populations in fluctuating environments: the case of the Serengeti wildebeest. *Journal of Applied Ecology* 32: 468–480.

Peterson, R. O. 1995. *The Wolves of Isle Royale–A Broken Balance*. Willow Creek Press, Minocqua, Wisconsin.

Peterson, R. O. 1999. Wolf-moose interactions on Isle Royale: the end of natural regulation? *Ecological Applications* 9: 10–16.

Pettorelli, N., J.-M. Gaillard, G. van Laere, et al. 2002. Variation in adult body mass in roe deer: the effects of population density at birth and of habitat quality. *Proceedings of the Royal Society B* 269: 747–753.

Pettorelli, N., F. Pelletier, A. von Hardenberg, M. Festa-Bianchet, and S. D. Côté. 2007. Early onset of vegetation growth vs. rapid green-up: impacts on juvenile mountain ungulates. *Ecology* 88: 381–390.

Portier, C., M. Festa-Bianchet, J.-M. Gaillard, J. T. Jorgenson, and N. G. Yoccoz. 1998. Effect of density and weather on survival of bighorn lambs. *Journal of Zoology* 245: 271–278.

Post, E., R. O. Peterson, N. C. Stenseth, and B. McLaren. 1999. Ecosystem consequences of wolf behavioural response to climate. *Nature* 401: 905–907.

Post, E. and N. C. Stenseth. 1998. Large-scale climatic fluctuation and population dynamics of moose and white-tailed deer. *Journal of Animal Ecology* 67: 537–543.

Ripple, W. J. and E. J. Larsen. 2000. Historic aspen recruitment, elk and wolves in northern Yellowstone National Park, USA. *Biological Conservation* 95: 361–370.

Romme, W. H., M. G. Turner, L. L. Wallace, and J. F. Walker. 1995. Aspen, elk, and fire in northern Yellowstone National Park. *Ecology* 76: 2097–2106.

Rose, K. E., T. H. Clutton-Brock, and F. E. Guinness. 1998. Cohort variation in male survival and lifetime breeding success in red deer. *Journal of Animal Ecology* 67: 979–986.

Sinclair, A. R. E. 1979. Dynamics of the Serengeti ecosystem, process and pattern. In *Serengeti: Dynamics of an Ecosystem*, eds. A. R. E. Sinclair and M. Norton-Griffiths, pp. 1–30. University of Chicago Press, Chicago.

Sinclair, A. R. E. and P. Arcese. 1995. Population consequences of predation-sensitive foraging: the Serengeti wildebeest. *Ecology* 76: 882–891.

Singer, F. J., A. Harting, K. K. Symonds, and M. B. Coughenour. 1997. Density dependence, compensation, and environmental effects on elk calf mortality in Yellowstone National Park. *Journal of Wildlife Management* 61: 12–25.

Smith, D. W., R. O. Peterson, and D. B. Houston. 2003. Yellowstone after wolves. *BioScience* 53: 330–340.

Taper, M. L. and P. J. P. Gogan. 2002. The northern Yellowstone elk: density dependence and climatic conditions. *Journal of Wildlife Management* 66: 106–122.

Vucetich, J. A. and R. O. Peterson. 2004. The influence of top-down, bottom-up and abiotic factors on the moose (*Alces alces*) population of Isle Royale. *Proceedings of the Royal Society B* 271: 183–189.

Vucetich, J. A., D. W. Smith, and D. R. Stahler. 2005. Influence of harvest, climate and wolf predation on Yellowstone elk, 1961–2004. *Oikos* 111: 259–270.

Wagner, F. 2006. *Yellowstone's Destabilized Ecosystems*. Oxford University Press, New York.

White, P. J. and R. A. Garrott. 2005. Yellowstone's ungulates after wolves – expectations, realizations and predictions. *Biological Conservation* 125: 141–152.

Wright, G. J., R. O. Peterson, D. W. Smith, and T. O. Lemke. 2006. Selection of northern Yellowstone elk by gray wolves and hunters. *Journal of Wildlife Management* 70: 1070–1078.

Zeigenfuss, L. C., F. J. Singer, S. A. Williams, and T. L. Johnson. 2002. Influences of herbivory and water on willow in elk winter range. *Journal of Wildlife Management* 66: 788–795.

2

The suite of population models

Norman Owen-Smith

School of Animal, Plant and Environmental Sciences, University of the Witwatersrand, Johannesburg, South Africa

Models provide the conceptual frameworks for understanding and explanation in ecology (Ford 2000). More specifically, "the utility of a model lies in its ability to produce accurate conceptual statements about specific natural systems" (Maurer 2004). Advances in scientific knowledge take place through "a cyclic process of model (re)construction and model (re)evaluation" (Taper and Lele 2004). A set of broadly adopted models constitutes the paradigm governing how a particular field of science is investigated.

In ecology, modeling has been approached from three largely disparate traditions (Hobbs 2009). Empiricists identify patterns evident in a particular data set and thereby obtain clues to the processes that might have generated these data. Statistical procedures are used to establish parameter values and the goodness of fit between the output projected by the model and the patterns seemingly apparent in the data. Theoreticians base their models on the biological processes that they assume to operate, expressed with mathematical elegance but with varying degrees of biological detail. At least vague phenomenological correspondence between the model predictions and real-world patterns helps reinforce beliefs in the validity of the underlying assumptions. Practical applications require more mechanistic detail, leading towards simulation models representing some structural framework and the connecting fluxes of materials or energy. Required parameter values may be assigned theoretically or derived from observations. Increasing sophistication in statistical

Dynamics of Large Herbivore Populations in Changing Environments, 1st edition. Edited by Norman Owen-Smith.
© 2010 Blackwell Publishing

procedures is opening the potential for a unification of these approaches (Burnham and Anderson 2002, Clark 2007, Cressie et al. 2009). All models simplify the real world in various ways, and exemplify the consequences of particular starting assumptions. Their validity depends on the extent to which the most important processes and connections have been identified.

Models that have been formulated to represent population dynamics have emphasized the following features:

1 *Density-dependence* regulating abundance around some equilibrium level (Sinclair 1989). Refinements draw on dynamical systems theory to express regulation in terms of a stationary probability distribution of abundance levels (Dennis and Taper 1994, Turchin 1995).
2 *Delayed feedbacks* promoting oscillations in abundance (Royama 1992, Turchin 2003). These can produce more complex dynamics, from dampened oscillations through limit cycles to chaos (May 1974a).
3 *Demographic structure* generating the population growth rate (Caswell 2001). This approach is largely concerned with the sensitivity of the growth rate to specific vital rates (Caswell 1978).
4 *Trophic interactions* with resources and predators (Crawley 1983, 1992). Delayed responses tend to generate reciprocal oscillations in abundance. These models may be elaborated to represent more complex food webs (May 1974b).
5 *Physiological* processes governing the conversion of energy and nutrients into population growth (Metz and Dieckmann 1986). The physiological mechanisms governing food extraction and conversion may be represented explicitly (Kooijman 1993, De Roos and Persson 2001) or at a more abstract *metaphysiological* level (Getz 1993).
6 *Individual-based* contributions to population growth, contingent on physiological and social state plus specific environmental contexts (DeAngelis and Gross 1992). Such models highlight emergent patterns arising from adaptive responses (Grimm and Railsback 2005).
7 *Landscape structure* affecting the population distribution (Bisonette 1997, Turner 2005). These models may encompass a distinction between source and sink subpopulations (Pulliam 1988).
8 *Metapopulation* units connected through dispersal (Hanski 1998, 1999), emphasizing processes of local extirpation and recolonization.

In the following sections, I review these modeling perspectives, and give examples where they have been applied to large mammal populations.

2.1 Models of density dependence

These models simply represent the effect of changing abundance on the population growth rate, leading to some abundance level or *carrying capacity* at which the net population growth becomes zero. The archetype is the logistic equation, expressing a linearly declining relationship between the proportional (or *relative*) growth rate and incremental changes in abundance:

$$dN/Ndt = r_0(1 - N/K)$$

where N represents population size (commonly expressed as numbers per unit area, or numerical density), r_0 the population growth rate towards very low abundance, and K the population size at which the growth rate becomes zero (Fig. 2.1a). The intrinsic growth potential r_0 is conceptualized as the maximum difference between overall birth and death rates.

However, population abundance is usually assessed at set time intervals, rather than continuously as implied by the differential equation formulation. This avoids the fact that numerical population increase generally takes place during a birth pulse within a restricted period of the annual cycle, followed by steady attrition as some of the new recruits as well as older animals die. Large mammal populations are most commonly censused once each year, either shortly after births have occurred or at some later, more convenient time. For comparison with these annual estimates of the population size, the logistic equation needs to be transformed into a discrete-time version, also called the *logistic map*:

$$(N_{t+1} - N_t)/N_t = r_0(1 - N_t/K) \tag{2.1}$$

where N_t represents the population size at time t. Notably, the relative population growth between one census time and the next now depends on the population size at the start of this interval. This discrete-time formulation subtly introduces a lag between cause and effect, because population growth depends on the abundance level one time step earlier without being influenced by subsequent changes in density. This delay generates dampened or persistent oscillations in abundance if the population increase is sufficiently large over the time step (May et al. 1974). Spurious indications of density dependence can also be generated when there are errors in the estimates of N_t (Shenk et al. 1998).

Nonlinear patterns of density dependence can be represented by incorporating an exponential transformation of the density-dependent factor, yielding the theta-logistic equation (Gilpin and Ayala 1973):

$$(N_{t+1} - N_t)/N_t = r_0\{1 - (N_t/K)^\theta\} \tag{2.2}$$

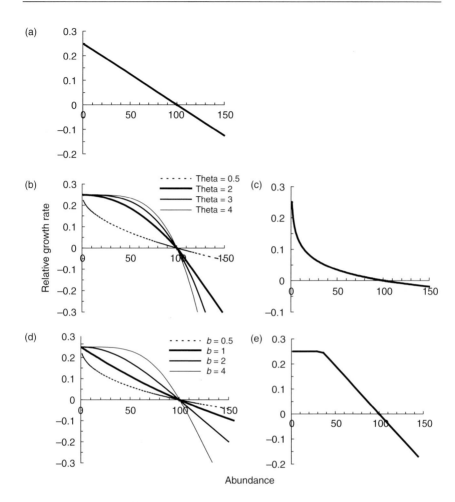

Figure 2.1 Forms of density dependence generated by simple equations. (a) Logistic equation. (b) Theta-logistic, for various values of θ. (c) Gompertz equation on an arithmetic scale. (d) Hyperbolic equation, for various values of b. (e) Threshold logistic equation.

Values of θ greater than 1 produce convex density dependence, while values less than 1 generate concave density dependence (Fig. 2.1b). The effective slope of the density relationship in the region of K is represented by the product $r_0\theta$ (Saether et al. 2000). Higher values of θ increase the extent of overshoot of K before the density feedback becomes effective one time step later. Convex density dependence can generate oscillations in abundance even for moderate values of the maximum growth rate (Getz 1996).

The subtraction term in Eqs. 2.1 and 2.2 can lead to negative numbers being generated for N. This inconvenience is avoided by using an exponential formulation of the logistic map, known as the *Ricker equation*:

$$N_{t+1} = N_t \exp \{r_{max}(1 - N_t/K)\}$$

where r_{max} representing the maximum population growth rate differs slightly from r_0. It is commonly applied using the exponential growth rate r_t, where $r_t = \log_e(N_{t+1}/N_t)$:

$$r_t = r_{max} (1 - N_t/K) \tag{2.3}$$

A θ-coefficient can also be incorporated into the Ricker equation to modify the form of density dependence:

$$r_t = r_{max} \{1 - (N_t/K)^\theta\} \tag{2.4}$$

Berryman (1999) labeled the functional relationship between the log-transformed population multiplication rate and population abundance the *R-function*.

A further modification entails a log-transformation of the abundance levels on both sides of the equation, producing the log-linear or Gompertz equation:

$$X_{t+1} = b_0 + (1 + b_1)X_t \tag{2.5}$$

where $X_t = \log_e(N_t)$, b_0 now represents the maximum exponential growth rate, and the coefficient b_1 is generally negative and expresses the strength of the density feedback. This formulation brings a statistical advantage, because the estimated value of N_t no longer appears on both sides of the equation, in the denominator for r_t as well as representing the density level producing the feedback on the population growth rate. However, the effectively proportional relationship with changing abundance generates a concave form of density dependence on an arithmetic scale (Fig. 2.1c).

Another expression, known as the *Beverton-Holt equation*, formulates a hyperbolic relationship between the population multiplication rate and population size or density:

$$N_{t+1}/N_t = R/(1 + aN_t) \tag{2.6}$$

where $R = 1 + r_0$, and the value of a expresses the strength of the density feedback. The zero growth level is not specified explicitly. Nevertheless, it is inherent in the joint values of R and a: $K = (R-1)/a$. This equation does not generate oscillations, no matter how large the value of R is, because increasing R accentuates the concavity in the density response, ensuring that the approach towards the zero growth level remains gradual. However, the form of the nonlinear density dependence can be varied by introducing a power coefficient b, producing a generalized sigmoid growth equation:

$$N_{t+1} = RN_t/(1 + [aN_t]^b) \qquad (2.7)$$

where the value of b determines the convexity or concavity in the density relationship (Fig. 2.1d). Setting values of b greater than 1 produces convex density dependence, and hence a susceptibility to oscillations in abundance (Getz 1996).

Lastly, an abrupt onset of density dependence above some threshold abundance could be represented by a piece-wise linear, or plateau-and-ramp, relationship (McCullough 1999; Fig. 2.1e). The steepness of the ramp, dependent on the threshold abundance at which density effects become manifested, controls the potential for oscillations to be generated. Moreover, the form of the density dependence above the threshold need not be linear.

The above equations are phenomenological in the sense that they merely describe the form of the functional relationship between rising abundance and the population growth rate without specifying the mechanistic basis. Nevertheless, a resource-based interpretation can be derived (Caughley 1976a, Berryman et al. 1995). If K, setting the zero growth level, represents the capacity of food resources to support the population, and N the aggregate population demand on this capacity, the ratio K/N is the effective food share obtained per individual. Hence, in the logistic model the relative growth rate is inversely dependent on this relative food share. In practice, other factors may hold the population below the ceiling abundance set by food resources. The density response is simply an outcome of unidentified factors associated with increased crowding.

Environmental variability can be accommodated through effects on either the value of K or r. Royama (1992) referred to influences affecting only the zero growth level K as lateral perturbations, and those affecting the realized growth rate r across all levels of abundance as vertical perturbations (Fig. 2.2). The extent of the fluctuations in abundance generated by variable K depends on the form of curvilinearity in the density response. Convex density dependence (θ or $b > 1$) amplifies oscillations, while concave (θ or $b < 1$) or log-linear density dependence suppresses them (Fig. 2.3).

Fowler (1981) concluded that convex density dependence was typical of large mammals, with the density feedback on population growth becoming expressed only towards high population levels. He interpreted this as an

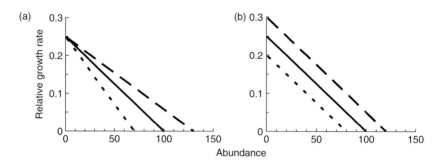

Figure 2.2 Potential effects of environmental variation on the relationship between population growth rate and abundance, as distinguished by Royama (1992). (a) Lateral perturbations, with the zero-growth level changed but not the inherent growth rate at low density. (b) Vertical perturbations, with the population growth rate altered across the entire density range.

outcome of the prevalence of resource limitations. Accordingly, the theta-logistic equation in arithmetic or exponential form is commonly adopted to represent the density influence for large herbivores. In projecting permissible harvests for elk in the National Elk Refuge, Wyoming, Boyce (1989) chose $\theta = 3.5$ as best fitting the recruitment relationship. For the dynamics of elk in northern Yellowstone, a simple Ricker equation with $\theta = 1$ was more strongly supported than the Gompertz model (Taper and Gogan 2002). For ibex in Swiss National Park, a Ricker equation with $\theta = 1.5$ was best supported, while log-linear density dependence represented by $\theta = 0$ fell outside the 95% confidence limits (Saether et al. 2002). For the population growth of elephants in Zimbabwe, the best-fit value of θ was 6.5, capturing the sharp onset of density dependence at high abundance (Chamaille-Jammes et al. 2008). For Soay sheep on Hirta, a setting of $b = 15$ in the generalized sigmoid growth equation best described the strongly nonlinear survival relationships, but with a modification to allow for the survival of some adult females even at high density (Grenfell et al. 1992). These findings confirm Fowler's (1981) impression that convex density dependence with a delayed onset is pervasive among large herbivore populations. The widespread prevalence of density dependence in ungulate populations was confirmed by the review of Bonenfant et al. (2009).

2.2   Autoregressive time-series models

Delayed as well as direct density feedbacks can be represented by expanding the population growth equation into an autoregressive time series

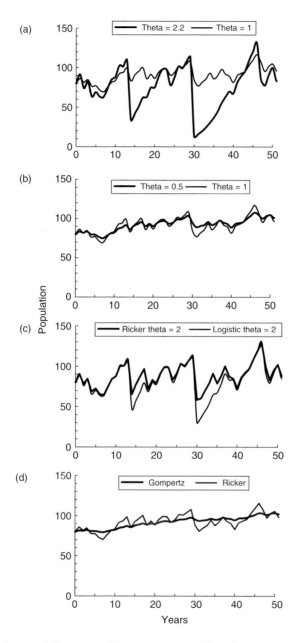

Figure 2.3 Comparative output dynamics generated by alternative formulations of the density feedback when K carrying capacity varies randomly over a range from 0.5 to 1.5 times its mean value. (a) Theta-logistic, $\theta = 2.2$ versus $\theta = 1$; (b) Theta-logistic, $\theta = 0.5$ versus $\theta = 1$; (c) Ricker, $\theta = 2$ versus logistic, $\theta = 2$; (d) Gompertz versus Ricker with $\theta = 1$.

(Royama 1992). The basic formulation takes the form

$$\log_e(N_t/N_{t-1}) = \beta_0 + \beta_1 N_{t-1} + \beta_2 N_{t-2} + \beta_3 N_{t-3} \dots \dots \quad (2.8)$$

where the βs represent the density coefficients for increasing time lags. This is essentially the Ricker equation with lagged terms added. The coefficient β_0 is commonly interpreted as representing the intrinsic rate of increase, while β_1 indexes the strength of direct density dependence, reducing the actual growth rate below this maximum. The ratio β_0/β_1 controls the effective value of K, assuming that the form of density dependence is linear. Delayed density feedbacks are interpreted as being mediated through biotic interactions with food resources and predators. The lag order of the model is indicated by the number of earlier abundance levels contributing to the dynamics (Turchin 2003). Depending on the relative strengths of the direct and delayed effects, a smooth or dampened approach towards equilibrium or sustained oscillations may be generated (Royama 1992; Fig. 2.4). Effects of weather and other exogenous factors on population growth are conventionally incorporated within a random error term ε added to the right-hand side of the equation. To avoid the statistical problem that arises when estimates of N_{t-1} are incorporated in both response and predictor terms, this equation is commonly rearranged into Gompertz form

$$X_t = \beta_0 + (1 + \beta_1)X_{t-1} + \beta_2 X_{t-2} + \beta_3 X_{t-3} \dots \dots \quad (2.9)$$

where $X_t = \log_e(N_t)$.

Relatively few models applied to large mammalian herbivores have considered delayed density feedbacks, on account of the limited duration of most time series of abundance for such long-lived species. For some red deer populations in Norway, Forchhammer et al. (1998) found indications of delayed density dependence from a year earlier than the 1-year lag automatically represented in the discrete-time formulation, operating through effects on the growth and fecundity of females. However, the magnitude seemed too weak to generate much variation in abundance (Loison and Langvatn 1998). No evidence of delayed density dependence was found for elk in northern Yellowstone (Taper and Gogan 2002). Turchin (2003) suggested that oscillations in the abundance of deer populations generated through lagged population interactions are likely to show a cycle period of 30–50 years, and thus only be evident in time series exceeding this duration. Oscillations in the abundance of certain ungulate species in Kruger National Park indicated cycle periods of 20–36 years, but appeared to be related to a cyclic component in the rainfall affecting

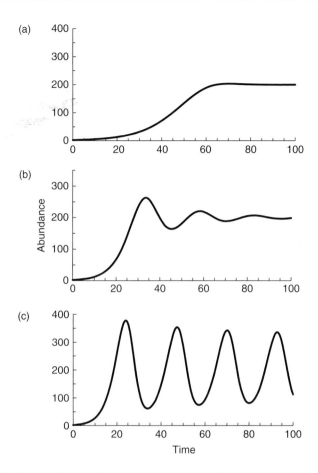

Figure 2.4 Effects of lagged density feedback on population trajectory as affected by the maximum population growth rate. (a) Lag order = 5 years, $r_m = 0.1$ per year. (b) Lag order = 5 years, $r_m = 0.2$ per year. (c) Lag order = 5 years, $r_m = 0.3$ per year.

susceptibility to predation rather than being intrinsically generated (Ogutu and Owen-Smith 2005).

2.3 Age- or stage-structured models

Demographic models project the overall population growth rate generated by a set of age- or stage-specific survival and fecundity rates (Caswell 2001). Estimates of these vital rates are obtained by following cohorts of identifiable individuals over time. Individuals are marked by some tag, and then "recaptured" at later times, provided they survive. Sometimes individuals can be recognized from variations in natural markings without placing tags. Sophisticated software has been developed for estimating these vital rates from capture-mark-recapture studies (Lebreton et al. 1992).

The multiplicative growth rate λ inherent in the set of vital rates may be ascertained using matrix algebra. Models typically consider only the female segment, assuming that males do not influence the fecundity rates. For an age-structured population, the Leslie matrix takes the form

$$
\begin{matrix}
0 & m_2 & m_3 & ..... & m_a \\
p_1 & 0 & 0 & .... & 0 \\
0 & p_2 & 0 & .... & 0 \\
0 & 0 & p_3 & .... & 0 \\
.... & .... & .... & .... & .... \\
0 & 0 & 0 & p_{a-1} & 0
\end{matrix}
$$

where m_x represents the specific fecundity of females in age class x, p_x represents the survival probability of individuals of starting age x into the next age class, and a is the maximum longevity. The dominant eigenvalue of this projection matrix yields the value of λ prevailing after the population structure has stabilized. At this stage, $N_{t+1} = \lambda N_t$, where N_{t+1} and N_t are vectors containing the numbers in different age or stage classes at successive time steps. All age classes are then increasing at the same rate λ. Matrix multiplication necessarily operates over discrete time steps.

Age-structured projection matrices are commonly used to compare the effects of altering specific vital rates on the overall population growth rate (Caswell 2000). The assessment can be prospective, projecting how much change in the population growth rate would occur if each vital rate was altered to the same degree, or retrospective, considering the actual variability manifested by each vital rate in the past. In a sensitivity analysis, each vital rate is altered by the same amount, e.g. an increase or decrease by 0.1. However, if baseline survival rates in the adult stage exceed 0.9, an upward adjustment by this amount is not feasible. In the corresponding elasticity analysis, each vital rate is adjusted by the same proportion, e.g. up or down by 25% of the mean. Again, an upward change by this magnitude is not possible for large ungulates with high adult survival rates. Hence, only the effects of lowering survival rates by the same amount or proportion in the adult stage can be compared with equivalent changes in juvenile survival or in fecundity rates. For large ungulates, changes in adult survival have a much greater effect on the population growth rate than corresponding changes in offspring survival, largely because the adult segment makes up a greater fraction of the population than

juveniles for long-lived species producing only a single young annually
(Gaillard et al. 1998). However, in practice, juvenile survival tends to be
more variable between years than adult survival rates, so that most of
the variation in year-to-year growth of ungulate populations is related
to changes in offspring recruitment. Nevertheless, when adult survival is
lowered, e.g. by hunting or predation, the effect on the population trend
is much greater than it would be from a similar amount of additional
mortality imposed on immature animals. Variable juvenile recruitment
means that, in practice, no stable age structure is attained (Tuljapurkar
1989).

The demographic rates contributing to overall population growth can
themselves be expressed as functions of the population density and environmental conditions, drawing on the equations used to express overall
population growth. Wilmers and Getz (2004) modified the generalized
hyperbolic equation (Eq. 2.7) to represent the survival rates of elk in Yellowstone, allowing for a modifying influence of snow depth on the density
relationship:

$$p_{a,t} = \{p_{\max}(\alpha)^b\}/\{(\alpha)^b + \delta_t(N_t)^b\} \qquad (2.10)$$

where $p_{a,t}$ is the survival probability of age class a at time t, $p_{\max}$ is
the baseline survival rate under ideal conditions, N_t is the population
biomass density, δ_t is the snow depth reducing food availability, α is the half
saturation parameter, and b is a shape parameter governing the abruptness
of the onset of density dependence (Fig. 2.5).

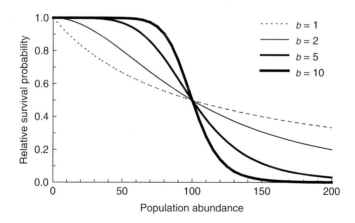

Figure 2.5 Relationship between the probability of survival relative to some maximum
value, and population density, generated by a modified hyperbolic equation for different
values of the shape parameter b (modified from Wilmers and Getz 2004).

For kudu in Kruger Park, where rainfall variation was the prime influence, a modification of the logistic equation was adopted:

$$p_{a,t} = p_{max} - a(N_{t-1}/k_t) \tag{2.11}$$

with the environmental conditions affecting the potential carrying capacity represented as a variable k_t, rather than a constant (Owen-Smith 1990, 2000). In this equation, k_t does not define the actual zero growth level, but only the strength of the feedback effect on survival at a particular density level. The abundance level at which population growth became zero would have to be found through simulation for particular combinations of the baseline survival rate and other parameter values.

Survival rates of particular age classes differentiated by year in the adult segment may, in practice, be indistinguishable, justifying their amalgamation into broader stage classes. However, for large ungulates it can be important to distinguish survival rates through some prime age range from those applying when advancing senescence leads to reductions in both survival and fecundity. Furthermore, the fecundity rates represented by m_x in the top row of the Leslie matrix are effectively the composite outcome of reductions in fecundity and survival of the offspring until the stage when the population is enumerated.

Under constant environmental conditions, linear relations between survival rates and population abundance reproduce the linear density response generated by the logistic model, apart from wobbles generated by changing age structure as the zero growth level is approached (Owen-Smith 2006). The threshold onset of density feedback depends on the varying density levels at which particular vital rates respond to the resource limitations associated with increasing abundance (Eberhardt 2002, Fig. 2.6). Hence, the overall form of the relationship between population growth and density dependence may appear convex (Owen-Smith 2006; Fig. 2.7a). A shift towards a greater proportion of older and hence more vulnerable animals when recruitment drops contributes to the negative density feedback (Festa-Bianchet et al. 2003). Lags in the age-structure adjustment tend to generate overshoot of the zero growth level, depending on the abruptness with which density effects take hold (Fig. 2.7b). The potential growth rate at low density may be truncated by additive mortality imposed by predation or other influences (Fig. 2.7c).

A demographic model based on stage-specific responses in survival to changing density and environmental conditions was developed for greater kudu in the Kruger National Park (Owen-Smith 1990, 2000). However, the population dynamics projected by this model under changing rainfall conditions were closely replicated by an equivalent logistic model assuming

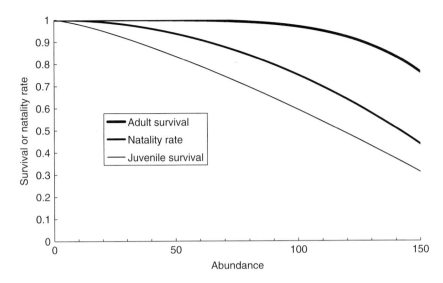

Figure 2.6 Comparative sensitivity of demographic contributions from juvenile survival, reproductive performance, and adult survival to rising density or other influences on resource availability, as proposed by Eberhardt (2002).

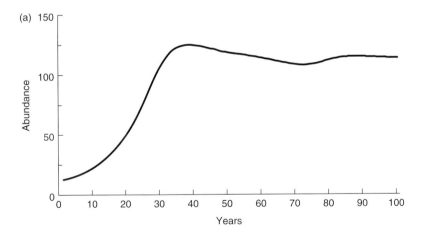

Figure 2.7 Effects of demographic structure on the population growth trajectory and form of density dependence. (a) Overshoot of the approach to the zero-growth level generated by adjustments in the age structure of a modeled white rhino population (from Owen-Smith 1988). (b) Overall density response generated by linear responses by different demographic segments becoming effective at different density levels. (c) Same as (b), but with population growth rate restricted by additive mortality towards low abundance levels (latter two figures from Owen-Smith 2006 for modeled kudu population).

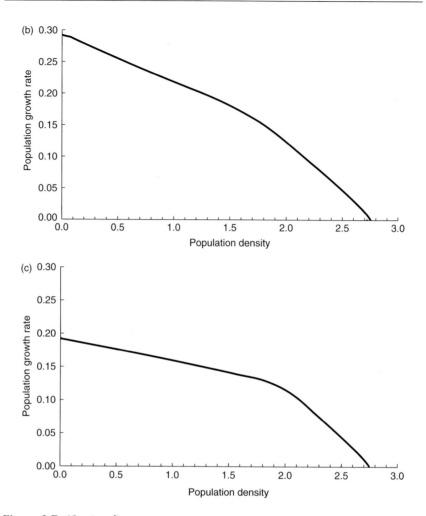

Figure 2.7 *(Continued)*

overall population growth to be a function of rainfall relative to density, without taking into account the population structure, after the appropriate values of r_{max} and K had been established. In contrast, for Soay sheep, an age-structured model produced a much better fit to the observed population fluctuations than simpler time-series models because of the widely varying sensitivity of different population segments to changing density and weather conditions (Coulson et al. 2001). In particular, this population became susceptible to crashes only when it contained a fairly high proportion of the juveniles and adult males that were most responsive in their survival rates to adverse conditions. The demographic approach can be resolved further towards explicit consideration of the changing hazards and risks to survival at different stages of the life cycle (Zens and Pearl 2003). A proportional hazards mortality model was applied to deer monitored by means of radiotelemetry by Fieberg and DelGuidice (2009).

2.4   Trophic interaction models

Trophic interaction models couple the dynamics of a consumer population to the reciprocal dynamics of its food resource (Lotka 1925, Volterra 1926). In general concept for a herbivore–plant interaction,

$$(1/H)dH/dt = G(V, H) - M(H)$$

and

$$(1/V)dV/dt = R(V) - I(V, H)$$

where H represents herbivore density or biomass and V, vegetation biomass (Caughley 1976a). R is the inherent production function for vegetation in the absence of herbivores, I is the functional (or intake) response of herbivores to changing vegetation availability, G is the gain function from food consumed into increase in herbivore abundance, and M represents intrinsic losses from the herbivore population. M can be interpreted as the rate at which the herbivore population would shrink in the absence of food. The difference G − M represents the so-called numerical response by the herbivore population to changing food availability.

Most commonly, the intake response is assumed to be hyperbolically saturating:
$$I = aV/(b + V)$$

where I = realized food intake rate per capita, a = maximum food intake rate, and b = amount of vegetation at which the food intake rate reaches half of its maximum. Accordingly,

$$(1/H)dH/dt = \{caV/(b + V)\} - m \qquad (2.12)$$

where c is the conversion coefficient from resource into consumer biomass, and m is the maintenance requirement, assumed to be constant. The herbivore population grows when nutritional gains exceed maintenance needs, and declines when gains fall short of this demand.

The maximum potential growth rate of the population can be projected by making the standing crop of vegetation V very large relative to the half-saturation level, so that food is not limiting. Accordingly

$$r_{max} = ca - m$$

Hence, the maximum growth rate depends on the rate at which herbivores can consume food and the efficiency with which they convert this food into their own biomass, relative to their maintenance requirements. The maximum population birth rate, relative to the death rate, is an expression of this potential. Restrictions in food availability reduce this rate, in particular the decline in standing vegetation biomass brought about as a result of consumption.

A logistic production function is commonly chosen for the vegetation biomass. Because of variable lags associated with the response of the consumer population to changing resource abundance, dependent on the parameter values, consumer dynamics may show either a smooth approach towards equilibrium abundance, dampened oscillations, or persistent cycling in abundance (Fig. 2.8a). Raising the productive potential of the vegetation in terms of the maximum standing crop that can be attained promotes greater instability (so-called "paradox of enrichment", Rosenzweig 1971). Elevating the conversion efficiency from food consumed into herbivore population increase has a similar effect.

In this basic model, the density feedback arises solely through the depletion of food as a consequence of consumption, reducing the rate of intake per unit herbivore. A hyperbolic intake response generates an asymptotically saturating population growth rate with increasing vegetation biomass (Fig. 2.8b). Below some threshold food requirement, the herbivore population growth becomes negative. Lags in the interaction with vegetation generate a spiralling trend in the population growth rate towards the equilibrium density level (Fig. 2.8c).

Competitive interference can be incorporated by adding a term dependent on the population abundance to the intake function, e.g.

$$I = aV/(b + V + \delta H)$$

where δ indexes the strength of competition (Beddington 1975, DeAngelis et al. 1975). In the extreme, the effective food availability can be represented by the resource share per unit herbivore, leading to the "ratio-dependent" formulation of the intake response (Arditi and Ginzburg 1989):

$$I = a(V/H)/\{b + (V/H)\}$$

The intake rate underlying population growth depends directly on the available vegetation V but inversely on the herbivore abundance H. Hence, the projected population growth rate is hyperbolically related to the herbivore biomass density if the amount of vegetation is held constant.

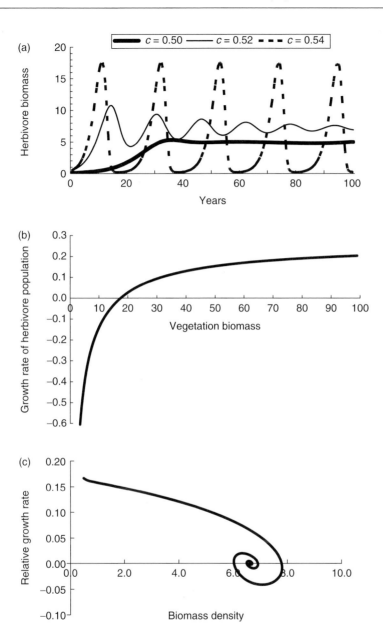

(a)

(b)

(c)

Figure 2.8 Output from coupled consumer–resource models. (a) Sensitivity of projected dynamics to small changes in the conversion coefficient c. (b) Dependence of population growth rate on available food biomass generated by a hyperbolic functional response. (c) Convex spiraling of the population growth rate towards zero generated by lags in the consumer–resource interaction.

Turchin (2003) proposed that the complex dynamics potentially generated by coupled consumer–resource interactions provides the most general foundation for population models, but applied this approach rather superficially to large ungulate populations. Unlike the situation for insects and small mammals, time lags arising through the population interactions may be somewhat longer than the annual time steps conventionally used in representing population changes. The conceptual herbivore–vegetation model developed by Caughley (1976a) was later parameterized to represent the dynamics of red kangaroo populations responding to wide fluctuations in their food resources in arid parts of Australia (Baylis 1987, Caughley 1987). However, the vegetation dynamics were simplified as a production pulse associated with high rainfall, subsequently depleted through consumption until critical thresholds leading to kangaroo mortality were surpassed. Choquenot (1998) used this approach to model the dynamics of feral pigs inhabiting riparian vegetation that was periodically flooded (see also Bayliss and Choquenot 2002). To project the dynamics of a white rhino population interacting with grasslands, I distinguished the seasonal reduction in food resources through consumption from long-term changes in the plant population producing this food (Chapter 13 in Owen-Smith 1988). Messier (1994) developed an interactive model incorporating the functional and numerical responses of wolves to changing moose abundance to assess the impact of this predator on moose population dynamics.

2.5   Physiological or metaphysiological models

Dynamic energy balance models represent population growth as being generated by rates of food acquisition and allocation to growth in size as well as reproduction, relative to metabolic and mortality losses (Kooijman 1993, Gutierrez 1996, De Roos and Persson 2001). Population structure is represented by size differences rather than age distinctions. A distinction is drawn between "irreversible" (structural) gains in individual size, and "reversible" contributions in the form of stored energy reserves. Metabolic costs are allometrically related to body size, while mortality losses depend on the ratio between reversible and irreversible body mass – when body reserves drop below some threshold proportion, death becomes certain.

Getz (1993, 1999) outlined a broader *metaphysiological* modeling approach to represent how resource fluxes influence population biomass dynamics within a food web (or "trophic stack") context. Emphasis was placed on the functional form of the relationships emerging from the underlying physiological processes. Both the intrinsic rate of population growth and the maximum density sustained can be derived as functions of the parameters governing resource fluxes. Instead of being represented as a constant loss, the mortality rate was expressed as inversely dependent on food gains. Stored reserves introduce a complication, effectively delaying the onset of mortality (Getz and Owen-Smith 1999). An outline of how such models enable resource-dependent foraging behavior to be related to the consequences for population dynamics is contained in Owen-Smith (2002).

Foraging models linked to energy balance, with consequences for survival and reproduction inferred, were developed for mule deer by Hobbs (1989) and for moose by Moen et al. (1997, 1998). For Soay sheep, Illius and Gordon (1998) related mortality to a critical threshold of fat reserves, and estimated the likely distribution of fat levels within the population as an integrated outcome of changing energy balance. Owen-Smith (2002, Chapter 13) applied the metaphysiological modeling approach to relate the oscillatory dynamics of Soay sheep to extreme seasonal food depletion, incorporating the differential sensitivity of life-history stages to food shortages. De Roos et al. (2009) applied a version of their dynamic energy balance model, taking into account age, size, and sex structure, to the dynamics of a feral horse population in Netherlands. Based on a threshold level of body condition that precipitates mortality, this model projected a series of years with mortality rates under 10%, interrupted by single years with overall mortality exceeding 40% during which almost all individuals under 2 years of age died.

2.6   Models accommodating spatial structure

Landscape models take into consideration the spatial distribution of organisms across regions differing in the resources, conditions, and deterrents to movement that they present (Turner et al. 1993, Ritchie 1997). They represent the contribution to local population dynamics of movements between habitats (Dunning et al. 1995), potentially including seasonal migrations (Fryxell et al. 1988). Source–sink versions differentiate subpopulations in favorable regions with a positive growth potential from those in less suitable habitat with a negative growth rate, counteracted by dispersal from the source to the sink region (Pulliam 1988). Metapopulation models represent the dynamics of discrete populations isolated by unsuitable habitat, through which animals disperse occasionally (Hanski 1999). They emphasize the likelihood that local populations will eventually become extirpated, leading towards regional species extinction unless recolonization rates exceed rates of local disappearance. The models may address the consequences of habitat fragmentation, transformation, and insularization for population viability.

Concepts relevant to large herbivores include key resource areas supporting animals through crucial periods in the seasonal cycle (Scoones 1995, Illius and O'Connor 1999) and habitat refuges providing security against predation and extreme weather (Berryman and Hawkins 2006). Geographic information systems allow broader distributional changes in response to climatic shifts together with other influences to be assessed. They enable niche dimensions dependent on habitat conditions as well as biotic interactions with resources, competitors, and predators to be identified (Guisan and Thuiller 2005, Soberon 2007).

2.7   Individual-based models

The finest structural decomposition of the population distinguishes the specific location, resource context, and social neighborhood of each individual constituting the population (Caswell and John 1992). Difficulties in parameterizing models at this resolution restrict applications of this "i-state configuration" approach. Nevertheless, consequences of individual distinctions can be explored using hypothetical parameter values. This perspective reveals how individual differences in susceptibility to mortality can provide a buffer against extreme changes in abundance (Lomnicki 1988). Individual-based models can also take into account finer details of spatial complexity in habitat patches and their juxtaposition (Mooij and DeAngelis 1998).

Opportunities to document environmental dependencies at individual animal level have been opened through developments in global positioning system (GPS) tracking technology. GPS tags documenting the sequential locations of a sample of animals according to some regular schedule reveal the changing space utilization patterns of these individuals and hence their local resource dependencies, in great detail (Kernohan et al. 2001). The challenge comes in scaling up observations on the individuals bearing the tracking collars to the wider population. Variations in space occupation have been documented for woodland caribou in British Columbia (Johnson et al. 2002), elk in Yellowstone National Park (Boyce et al. 2003), and sable antelope in Kruger National Park (Owen-Smith and Cain 2007). A model has been developed relating the extent of the home range and the degree to which it is shared to the population abundance level attained, but it has not yet been applied to ungulates (Moorcroft and Barnett 2008).

2.8   Overview

The models that have been most widely applied emphasize density-dependent feedbacks regulating populations around some equilibrium carrying capacity, rather than the environmental influences causing populations to depart from any fixed abundance level – sometimes quite widely. Convex forms of density dependence in population growth rate tend to promote oscillations, if feedbacks are delayed relative to the response time, but processes underlying the threshold onset of density dependence and the lags in response have not been adequately identified. The convexity may arise simply from the different thresholds at which particular demographic segments respond to diminishing resource availability, without the trend necessarily being nonlinear through the zero-growth level (Owen-Smith 2006). Delayed density dependence may arise demographically through age structure adjustments or cohort effects, or emanate from lags in population responses to changing resource availability. The latter delays may extend over several years rather than matching the annual time step conventionally used in discrete-time models.

The effects of weather variation are commonly interpreted in time-series models as stochastic "noise" disrupting populations from some equilibrium state (Bjornstad and Grenfell 2001). Extreme weather conditions can be represented as either lowering the maximum possible growth rate r_{max} or the potential carrying capacity K, but in reality, they simply cause elevated mortality. Nevertheless, this mortality may result either because less food is available as a result of the adverse weather conditions (droughts in African savannas, snow crusts in high northern latitudes), or because of the thermal stress imposed by cold, wet, and windy conditions. Vital rates can be expressed as functions of food production relative to the population demand, but matrix algebra becomes problematic if these rates do not remain constant.

Models depicting coupled consumer–resource dynamics represent the resource fluxes underlying changes in survival and reproduction, but do not represent the multiple populations that effectively constitute the food resource, nor restrictions on what fraction of the resource population biomass is effectively available for consumption. Incorporating some ungrazable amount generating regrowth but not exposed to consumption drastically changes the propensity of herbivore–vegetation models to oscillations (Turchin and Batzli 2001). Adding predation to the plant–herbivore interaction can also suppress the oscillatory tendency (Caughley 1976a). However, such models are vague about the time frames they represent and whether population abundance is expressed numerically or in terms of biomass density. The former may be adequate for predator–prey models, but counting individual plants is meaningless. These models generally do not distinguish seasonal variation in the production and continuing availability of vegetation resources from annual changes in the plant populations producing this potential food for herbivores. Furthermore, the disrupting or entraining effects of weather patterns on population dynamics are not considered.

Physiologically structured population models represent the consequences of energy and material gains for growth, reproduction, and survival in greater detail, but give less attention to the changing environmental conditions affecting these fluxes. Individual-based models open the potential to incorporate adaptive behavior in response to changing conditions. GPS tracking of animal movements opens possibilities to parameterize space use patterns at fine resolution, but links with population dynamics remain generally undeveloped.

Most theory in population ecology has been developed with small mammal or insect populations in mind, which are more readily amenable to experimentation and multigeneration observations than large, long-lived ungulates. The emphasis has shifted from explaining processes stabilizing dynamics towards interpreting those generating more or less regular fluctuations in abundance (Royama 1992, Berryman 1999, Turchin 2003). Counterbalancing this are avian studies emphasizing the stabilizing effects of social spacing mechanisms during the breeding season plus mortality incurred from unknown causes following post-breeding dispersal (Wynne-Edwards 1962, Lack 1966). Saether et al. (2004) specifically distinguished the consequences of weather variation during the breeding season affecting recruitment (the "tap" mechanism) from those occurring during the nonbreeding period affecting mortality ("tub" mechanism). For fish populations, the emphasis has been on wide recruitment variability in the context of the sustainability of harvest quotas (Walters and Martell 2004).

It is evident from Chapter 1 that large herbivore populations can show a wide range of dynamics, but the putative mechanisms underlying these contrasting patterns need to be explored more widely to establish their generality. More case studies are needed of situations in which large herbivore populations have increased to abundance levels threatening the persistence of vegetation resources, regarded as a prevalent problem in the management of deer in North America (McShea et al. 1997) and elephants in southern Africa (Caughley 1976b, Owen-Smith et al. 2006). Not represented are detailed studies of ungulate populations declining towards levels threatening their viability (Ogutu and Owen-Smith 2003, Wittmer et al. 2005), demanding diagnosis of the underlying causal mechanisms.

The chapters that follow explore the environmental influences governing changes in population abundance in greater detail, highlighting the interplay between model expectations and empirical findings. Chapter 3 addresses the manifold effects of climatic variation, as well as regional distinctions in predation and hunting pressure, on population trends, elaborating time-series models. Chapter 4 considers how climatic influences interact with the demographic structure and specific vital rates to affect the emergent population dynamics. Chapter 5 outlines the interplay between irruptive population growth and periodic crashes in abundance mediated through herbivore–vegetation interactions in the context of climatic variation. Chapter 6 explains the contribution of spatial variation in resource quality as well as the spatially changing availability of these resources towards population dynamics. Finally, Chapter 7 considers the further advances in modeling frameworks and paradigms needed to better represent the consequences of environmental variation in space and through time for population abundance levels.

Acknowledgments

I thank Bruce Kendall for the constructive comments on an earlier draft of this chapter, and John Fryxell for suggesting further improvements.

References

Arditi, R. and L. R. Ginzburg. 1989. Coupling in predator-prey dynamics: ratio dependence. *Journal of Theoretical Biology* 139: 311–326.

Bayliss, P. 1987. Kangaroo dynamics. In *Kangaroos: Their Ecology and Management in the Sheep Rangelands of Australia*, eds. G. Caughley, N. Shepherd, and J. Short, pp. 119–134. Cambridge University Press, Cambridge.

Bayliss, P. and D. Choquenot. 2002. The numerical response: rate of increase and food limitation in herbivores and predators. *Philosophical Transactions of the Royal Society of London B* 357: 1233–1248.

Beddington, J. R. 1975. Mutual interference between parasites or predators and its effects on searching efficiency. *Journal of Animal Ecology* 44: 331–340.

Berryman, A. A. 1999. *Principles of Population Dynamics and their Application*. Stanley Thornes, Cheltenham.

Berryman, A. A. and B. A. Hawkins. 2006. The refuge as an integrating concept in ecology and evolution. *Oikos* 115: 192–196.

Berryman, A. A., J. Michalski, A. P. Gutierrez, and R. Arditi. 1995. Logistic theory of food web dynamics. *Ecology* 76: 336–343.

Bisonette, J. (ed.) 1997. *Wildlife and Landscape Ecology: Effects of Pattern and Scale*. Springer, New York.

Bjornstad, O. N. and B. T. Grenfell. 2001. Noisy clockwork: time series analysis of population fluctuations in animals. *Science* 293: 638–643.

Bonenfant, C., J.-M. Gaillard, L. E. Loe, et al. 2009. Empirical evidence of density dependence in populations of large herbivores. *Advances in Ecological Research* 41: 313–357.

Boyce, M. S. 1989. *The Jackson Elk Herd. Intensive Wildlife Management in North America*. Cambridge University Press, Cambridge, UK.

Boyce, M. S., J. S. Mao, E. H. Merrill, et al. 2003. Scale and heterogeneity in habitat selection by elk in Yellowstone National Park. *Ecoscience* 10: 421–431.

Burnham, K. P. and D. R. Anderson. 2002. *Model Selection and Multi-Model Inference: A Practical Information-Theoretic Approach.* 2nd edition. Springer-Verlag, New York.

Caswell, H. 1978. A general formula for the sensitivity of population growth rate to changes in life history parameters. *Theoretical Population Biology* 14: 215–230.

Caswell, H. 2000. Prospective and retrospective perturbation analyses and their use in conservation biology. *Ecology* 81: 619–627.

Caswell, H. 2001. *Matrix Population Models. Construction, Analysis, and Interpretation.* 2nd edition. Sinauer, Sunderland, Massachusetts.

Caswell, H. and A. M. John. 1992. From the individual to the population in demographic models. In *Individual-based Models and Approaches in Ecology*, eds. D. L. DeAngelis and L. J. Gross, pp. 36–61. Chapman and Hall, New York.

Caughley, G. 1976a. Plant-herbivore systems. In *Theoretical Ecology*, ed. R. M. May, pp. 94–113. Blackwell, Oxford.

Caughley, G. 1976b. The elephant problem - an alternative hypothesis. *East African Wildlife Journal* 14: 265–283.

Caughley, G. 1987. Ecological relationships. In *Kangaroos: Their Ecology and Management in the Sheep Rangelands of Australia*, eds. G. Caughley, N. Shepherd, and J. Short, pp. 159–187. Cambridge University Press, Cambridge.

Chamaille-Jammes, S., H. Fritz, M. Valeix, F. Murindagomo, and J. Clobert. 2008. Resource availability, aggregation and direct density dependence in a open context: the local regulation of an African elephant population. *Journal of Animal Ecology* 77: 135–144.

Choquenot, D. 1998. Testing the relative influence of intrinsic and extrinsic variation in food availability on feral pig populations in Australia's rangelands. *Journal of Animal Ecology* 67: 887–907.

Clark, J. S. 2007. *Models for Ecological Data.* Princeton University Press, Princeton.

Coulson, T., E. A. Catchpole, S. D. Albon, et al. 2001. Age, sex, density, winter weather, and population crashes in Soay sheep. *Science* 292: 1528–1531.

Crawley, M. J. 1983. *Herbivory. The Dynamics of Animal-Plant Interactions.* Blackwell, Oxford.

Crawley, M. J. 1992. *Natural Enemies. The Population Biology of Predators, Parasites and Diseases.* Blackwell, Oxford.

Cressie, N., C. A. Calder, J. S. Clark, J. M. Ver Hoef, and C. K. Wikle. 2009. Accounting for uncertainty in ecological analysis: the strengths and limitations of hierarchical statistical modeling. *Ecological Applications* 19: 553–570.

DeAngelis, D. L., R. A. Goldstein, and R. V. O'Neil. 1975. A model for trophic interactions. *Ecology* 56: 881–892.

DeAngelis, D. L. and L. J. Gross (eds). 1992. *Individual-Based Models and Approaches in Ecology. Populations, Communities and Ecosystems.* Chapman & Hall, London.

Dennis, B. and M. L. Taper. 1994. Density-dependence in time series observations of natural populations: estimation and testing. *Ecological Monographs* 64: 205–224.

De Roos, A. M., N. Galic, and H. Heesterbeek. 2009. How resource competition shapes individual life history for non-plastic growth: ungulates in seasonal food environments. *Ecology* 90: 945–960.

De Roos, A. M. and L. Persson. 2001. Physiologically structured models - from versatile technique to ecological theory. *Oikos* 94: 51–71.

Dunning, J. B., D. J. Stewart, B. J. Danielson, et al. 1995. Spatially explicit population models: current forms and future uses. *Ecological Applications* 5: 3–11.

Eberhardt, L. L. 2002. A paradigm for population analysis of long-lived vertebrates. *Ecology* 83: 2841–2854.

Festa-Bianchet, M., J.-M. Gaillard, and S. D. Cote. 2003. Variable age structure and apparent density dependence in survival of adult ungulates. *Journal of Animal Ecology* 72: 640–649.

Fieberg, J. and G. D. DelGuidice. 2009. What time is it? Choice of time origin and scale in extended proportional hazards models. *Ecology* 90: 1687–1697.

Forchhammer, M., N. C. Stenseth, E. Post, and R. Langvatn. 1998. Population dynamics of Norwegian red deer: density dependence and climatic variation. *Proceedings of the Royal Society of London B* 265: 341–350.

Ford, E. D. 2000. *Scientific Method for Ecological Research.* Cambridge University Press, Cambridge.

Fowler, C. W. 1981. Density dependence as related to life history strategy. *Ecology* 62: 602–610.

Fryxell, J., J. Greever, and A. R. E. Sinclair. 1988. Why are migratory ungulates so abundant? *American Naturalist* 131: 781–798.

Gaillard, J.-M., M. Festa-Bianchet, and N. G. Yoccoz. 1998. Population dynamics of large herbivores: variable recruitment with constant adult survival. *Trends in Ecology and Evolution* 13: 58–63.

Getz, W. M. 1993. Metaphysiological and evolutionary dynamics of populations exploiting constant and interactive resources: r-K selection revisited. *Evolutionary Ecology* 7: 287–305.

Getz, W. M. 1996. A hypothesis regarding the abruptness of density dependence and the growth rate of populations. *Ecology* 77: 2014–2026.

Getz, W. M. 1999. Population and evolutionary dynamics of consumer-resource systems. In *Advanced Ecological Theory*, ed. J. McGlade, pp. 194–231. Blackwell, Oxford.

Getz, W. M. and N. Owen-Smith. 1999. A metaphysiological population model of storage in variable environments. *Natural Resource Modeling* 12: 197–230.

Gilpin, M. E. and F. J. Ayala. 1973. Global models of growth and competition. *Proceedings of the National Academy of Science of the USA* 70: 3590–3593.

Grenfell, B. T., O. F. Price, S. D. Albon, and T. H. Clutton-Brock. 1992. Overcompensation and population cycles in an ungulate. *Nature* 355: 823–826.

Grimm, V. and S. F. Railsback. 2005. *Individual-based Modeling and Ecology.* Princeton University Press, Princeton, New Jersey.

Guisan, A. and W. Thuiller. 2005. Predicting species distribution: offering more than simple habitat models. *Ecology Letters* 8: 993–1009.

Gutierrez, A. P. 1996. *Applied Population Ecology: A Supply-Demand Approach.* John Wiley, New York.

Hanski, I. 1998. Metapopulation ecology. *Nature* 396: 41–49.

Hanski, I. 1999. *Metapopulation Ecology.* Oxford University Press, Oxford.

Hobbs, N. T. 1989. Linking Energy Balance to Survival in Mule Deer: Development and Test of a Simulation Model. *Wildlife Monographs* No. 101.

Hobbs, N. T. 2009. New tools for insight from ecological models and data. *Ecological Applications* 19: 551–552.

Illius, A. W. and I. J. Gordon. 1998. Scaling up from functional response to numerical response in vertebrate herbivores. In *Herbivores, Plants and Predators*, eds. H. Olff, V. K. Brown, and R. T. Drent, pp. 397–427. Blackwell, Oxford.

Illius, A. W. and T. G. O'Connor. 1999. On the relevance of nonequilibrium concepts to arid and semi-arid grazing systems. *Ecological Applications* 9: 798–813.

Johnson, C. J., K. L. Parker, D. C. Heard, and M. P. Gillingham. 2002. A multi-scale behavioral approach to understanding the movements of woodland caribou. *Ecological Applications* 12: 1840–1860.

Kernohan, B. J., R. A. Gitzen, and J. J. Millspaugh. 2001. Analysis of animal space use and movements. In *Radiotracking and Animal Populations*, eds. J. J. Millspaugh and J. M. Marzluff, pp. 126–166. Academic Press, San Diego.

Kooijman, S. A. L. M. 1993. *Dynamic Energy Budgets in Biological Systems*. Cambridge University Press, Cambridge.

Lack, D. 1966. *Population Studies of Birds*. Oxford University Press, Oxford.

Lebreton, J.-D., K. P. Burnham, J. Clovert, and D. R. Anderson. 1992. Modeling survival and testing biological hypotheses using marked animals: a unified approach with case studies. *Ecological Monographs* 62: 67–118.

Loison, A. and R. Langvatn. 1998. Short- and long-term effects of winter and spring weather on growth and survival of red deer in Norway. *Oecologia* 116: 489–500.

Lomnicki, A. 1988. *Population Ecology of Individuals*. Princeton University Press, Princeton.

Lotka, A. J. 1925. *Elements of Mathematical Biology*. Dover, New York.

Maurer, B. A. 2004. Models of scientific inquiry and statistical practice: implications for the structure of scientific knowledge. In *The Nature of Scientific Evidence*, eds. M. L. Taper and S. R. Lele, pp. 17–31. University of Chicago Press, Chicago.

May, R. M. 1974a. Biological populations with non-overlapping generations: stable points, stable cycles and chaos. *Science* 186: 645–647.

May, R. M. 1974b. *Stability and Complexity in Model Ecosystems*. 2nd edition. Princeton University Press, Princeton.

May, R. M., G. R. Conway, M. P. Hassell, and T. R. E. Southwood. 1974. Time delays, density dependence and single species oscillations. *Journal of Animal Ecology* 43: 747–770.

McCullough, D. R. 1999. Density dependence and life-history strategies of ungulates. *Journal of Mammalogy* 80: 1130–1146.

McShea, W. J., H. B. Underwood, and J. H. Rappole (eds). 1997. *The Science of Overabundance: Deer Ecology and Population Management*. Smithsonian Institute Press, Washington, DC.

Messier, F. 1994. Ungulate population models with predation: a case study with the North American moose. *Ecology* 75: 478–488.

Metz, J. A. J. and O. Dieckmann. 1986. *The Dynamics of Physiologically Structured Populations*. Springer, Berlin-Hamburg.

Moen, R., Y. Cohen, and J. Pastor. 1998. Linking moose population and plant growth models with a moose energetics model. *Ecosystems* 1: 52–63.

Moen, R., J. Pastor, and Y. Cohen. 1997. A spatially explicit model of moose foraging and energetics. *Ecology* 78: 505–521.

Mooij, W. M. and D. L. DeAngelis. 1998. Individual-based modelling as an integrative approach in theoretical and applied population dynamics and food web studies. In *Herbivores, Plants and Predators*, eds. H. Olff, V. K. Brown, and R. T. Drent, pp. 551–575. Blackwell, Oxford.

Moorcroft, P. R. and A. Barnett. 2008. Mechanistic home range models and resource selection analysis: a reconciliation and unification. *Ecology* 89: 1112–1119.

Ogutu, J. O. and N. Owen-Smith. 2003. ENSO, rainfall and temperature influences on extreme population declines among African savanna ungulates. *Ecology Letters* 6: 412–419.

Ogutu, J. O. and N. Owen-Smith. 2005. Oscillations in large herbivore populations: are they related to predation or rainfall? *African Journal of Ecology* 43: 332–339.

Owen-Smith, N. 1988. *Megaherbivores. The Influence of Very Large Body Size on Ecology.* Cambridge University Press, Cambridge.

Owen-Smith, N. 1990. Demography of a large herbivore, the greater kudu, in relation to rainfall. *Journal of Animal Ecology* 59: 893–913.

Owen-Smith, N. 2000. Modeling the population dynamics of a subtropical ungulate in a variable environment: rain, cold and predators. *Natural Resource Modeling* 13: 57–87.

Owen-Smith, N. 2002. *Adaptive Herbivore Ecology. From Resources to Populations in Variable Environments.* Cambridge University Press, Cambridge.

Owen-Smith, N. 2006. Demographic determination of the shape of density dependence for three African ungulate populations. *Ecological Monographs* 76: 73–92.

Owen-Smith, N. and J. W. Cain III. 2007. Indicators of adaptive responses in home range utilization and movement patterns by a large mammalian herbivore. *Israel Journal of Ecology and Evolution* 53: 423–438.

Owen-Smith, N., G. I. H. Kerley, B. Page, R. Slotow and R. J. Van Aarde. 2006. A scientific perspective on the management of elephants in the Kruger National Park and elsewhere. *South African Journal of Science* 102: 389–394.

Pulliam, H. R. 1988. Sources, sinks, and population regulation. *American Naturalist* 137: S50–S66.

Ritchie, M. E. 1997. Populations in a landscape context: sources, sinks and metapopulations. In *Wildlife and Landscape Ecology*, ed. J. A. Bissonette, pp. 160–184. Springer, New York.

Rosenzweig, M. L. 1971. Paradox of enrichment: destabilization of exploitation ecosystems in ecological time. *Science* 171: 385–387.

Royama, T. 1992. *Analytical Population Dynamics.* Chapman & Hall, London.

Saether, B. E., S. Engen, F. Filli, R. Aanes, W. Schroder, and R. Andersen. 2002. Stochastic population dynamics of an introduced Swiss population of the ibex. *Ecology* 83: 3457–3465.

Saether, B. E., S. Engen, R. Lande, P. Arcese, and J. N. M. Smith. 2000. Estimating the time to extinction in an island population of song sparrows. *Proceedings of the Royal Society of London, Series B* 267: 621–626.

Saether, B. -E., W. J. Sutherland, and S. Engen. 2004. Climate influences on avian population dynamics. *Advances in Ecological Research* 35: 185–209.

Scoones, I. 1995. Exploiting heterogeneity: habitat use by cattle in dryland Zimbabwe. *Journal of Arid Environments* 29: 221–237.

Shenk, T. M., G. C. White, and K. P. Burnham. 1998. Sampling-variance effects on detecting density dependence from temporal trends in natural populations. *Ecological Monographs* 68: 445–463.

Sinclair, A. R. E. 1989. Population regulation in animals. In *Ecological Concepts.* 29th Symposium of the British Ecological Society, ed. J. M. Cherrett, pp. 197–241. Blackwell, Oxford.

Soberon, J. 2007. Grinnellian and Eltonian niches and geographic distributions of species. *Ecology Letters* 10: 1115–1123.

Taper, M. L. and P. J. P. Gogan. 2002. The northern Yellowstone elk: density dependence and climatic conditions. *Journal of Wildlife Management* 66: 106–122.

Taper, M. L. and S. R. Lele. 2004. Dynamical models as paths to evidence in ecology. In *The Nature of Scientific Evidence*, eds. M. L. Taper and S. R. Lele, pp. 275–297. University of Chicago Press, Chicago.

Tuljapurkar, S. 1989. An uncertain life: demography in random environments. *Theoretical Population Biology* 35: 227–294.

Turchin, P. 1995. Population regulation: old arguments and a new synthesis. In *Population Dynamics*, eds. N. Cappuccino and P. W. Price, pp. 19–40. Academic Press, San Diego.

Turchin, P. 2003. *Complex Population Dynamics. A Theoretical/Empirical Synthesis.* Princeton University Press, Princeton, NJ.

Turchin, P. and G. Batzli. 2001. Availability of food and the population dynamics of arvicolline rodents. *Ecology* 82: 1521–1534.

Turner, M. G. 2005. Landscape ecology in North America: past, present and future. *Ecology* 86: 1967–1974.

Turner, M. G., Y. Wu, W. H. Romme, and L. L. Wallace. 1993. A landscape simulation model of winter foraging by large ungulates. *Ecological Modelling* 69: 163–184.

Volterra, V. 1926. Fluctuations in the abundance of species considered mathematically. *Nature* 118: 558–560.

Walters, C. J. and J. D. Martell. 2004. *Fisheries Ecology and Management.* Princeton University Press, Princeton, NJ.

Wilmers, C. C. and W. M. Getz. 2004. Simulating the effects of wolf-elk population dynamics on resource flow to scavengers. *Ecological Modelling* 177: 193–208.

Wittmer, H. U., B. N. McLellan, D. R. Seip, et al. 2005. Population dynamics of the endangered mountain ecotype of woodland caribou in British Columbia, Canada. *Canadian Journal of Zoology* 83: 407–418.

Wynne-Edwards, V. C. 1962. *Animal Dispersion in Relation to Social Behaviour.* Oliver & Boyd, Edinburgh.

Zens, M. S. and D. R. Pearl. 2003. Dealing with death data: individual hazards, mortality and bias. *Trends in Ecology and Evolution* 18: 366–373.

3

Climatic influences: temperate–tropical contrasts

Norman Owen-Smith

School of Animal, Plant and Environmental Sciences, University of the Witwatersrand, Johannesburg, South Africa

Anticipating the consequences of global climate change constitutes one of the major scientific challenges for the 21st century. Chesson (2003) noted that although temporal fluctuations in the physical environment are recognized to be a major driver of population fluctuations, they have received relatively scant theoretical attention in the past. Environmental variation is commonly relegated to the status of stochastic "noise," disrupting the population trend towards some equilibrium density. Operationally, however, variable weather is the primary driver of changes in population abundance, with the prevailing density level merely modifying its expression.

Nevertheless, relationships between climatic factors and population dynamics can be complex and often indirect, and the challenge is how to disentangle the causal links. The focus of this chapter is on how climatic variation influences changes in large herbivore abundance, in interaction with the density level, resource availability, and predation or hunting pressure. Climate and its expression through local weather conditions may affect large herbivores directly, by imposing mortality through cold or heat stress, or indirectly via the production of food and its seasonal availability. Populations respond to weather variation over seasonal, annual, and multi-annual time frames, and there can be carryover effects from past conditions. In temperate latitudes, the seasonal cycle in plant growth is controlled largely by temperature variation, and is fairly predictable. In seasonally dry tropical and subtropical latitudes, the controlling rainfall

Dynamics of Large Herbivore Populations in Changing Environments, 1st edition. Edited by Norman Owen-Smith.
© 2010 Blackwell Publishing

influence is more erratic, so that food availability at a particular time of the year may be somewhat less predictable.

Predators may alter the effects of weather, by making animals vulnerable through malnutrition or hampered in their movements by snow, and thus making them more likely to die. However, if predation holds populations somewhat below the abundance level where food becomes limiting, animals may be able to resist extreme weather conditions by drawing on stored body reserves. Similarly, populations that have their abundance controlled by hunting or management culling may be less exposed to weather influences than those limited ultimately by food resources in the absence of much hunting or predation. A full complement of large mammalian predators generally persists within the tropical or subtropical savanna environments of Africa, where large herbivore populations are generally studied. In contrast, within the more temperate environments of Europe and North America, large carnivores have been mostly eliminated, but widely replaced by human hunters. Hence, distinctions in the dynamics of large herbivore populations between these two broad regions that might be ascribed to climatic contrasts could simply be an outcome of differences in predation or hunting pressure.

The weather aspects affecting population dynamics potentially include temperature extremes, wind force, and precipitation in the form of rain or snow. Their effects depend on the time of year when they occur, how long extreme conditions persist, and on other circumstances. Similar variation in weather may have different effects under different conditions, e.g. elevated temperatures in early spring extend the growing season, potentially increasing food production, while elevated temperatures during late winter can lead to ice crusts forming on snow, blocking access to the forage buried beneath. Including too many weather factors individually in models may overwhelm the capacity of the data to discriminate their effects. Hence it can be advantageous to capture the effects of a set of weather components that covary (i.e. a weather package) by means of some index of the atmospheric conditions generating them (Stenseth et al. 2003, Stenseth and Mysterud 2005).

Weather conditions variation may disrupt, obscure, or reinforce the density feedbacks that counteract excessively high or insecurely low population levels. Establishing the functional form of the density response requires factoring out these other influences on the population growth rate.

Correlative models represent a first step towards the predictive capability needed to guide management decisions to counteract or ameliorate the consequences of climate shifts (Krebs and Berteaux 2006). Weather influences can be assessed by modifying autoregressive time-series models to include additional factors besides current or prior density levels:

$$r_t = \beta_0 + \beta_1 Y_1 + \beta_2 Y_2 + \beta_3 Y_3 \ldots \ldots \ldots + \varepsilon \qquad (3.1)$$

where $r_t = \log_e(N_{t+1}/N_t)$, the Ys represent the candidate set of factors, the βs index their effects on population growth, expressed in simple linear terms, and ε is the error component unexplained by the factors included in the model plus sampling error. Besides representing the abundance levels at earlier time steps, some of the Ys need to be equated with particular weather factors. Weather conditions may either have an independent additive effect on population growth, or modify the influence of current or earlier density levels. The effects of weather may also be lagged via vegetation changes induced, and the consequences of extreme conditions may persist for several years.

In this chapter, I review the evidence indicating how weather features have affected the growth rate of large herbivore populations, and outline the models that have been formulated to represent these influences. The chapter is structured into two sections–the first addressing patterns observed in temperate or higher northern latitudes, and the second considering findings in tropical or subtropical environments, largely in Africa. Within each section, I consider first the effects of individual weather components, and thereafter the ability of aggregate indices of weather packages to better capture these effects. I also assess how predation or hunting might have affected the influence of weather, as well as the modifying influence of the population abundance. The questions to be borne in mind are as follows:

1 Which weather measures should be represented?
2 What form does their influence take–linear, curvilinear, or threshold?
3 Is their effect direct, or mediated through food resources?
4 How is their effect modified by the prevailing population density level?
5 How does predation or hunting affect the response to weather variation?

3.1 Temperate environments

3.1.1 Individual weather components

In high northern latitudes, temperature variation largely controls the seasonal cycle in plant growth. Precipitation is less influential, because soil moisture is retained through cold winters, while snowmelt reliably facilitates spring growth. However, precipitation accumulating as snow during winter makes food less accessible for herbivores dependent on herbaceous plants, and deep snow may delay the exposure of plants to rising temperatures in spring. Towards less extreme latitudes and elevations, snow cover becomes less of a factor, and precipitation in the form of rain during summer has a greater influence on vegetation growth.

Warmer conditions at the end of winter promote an earlier onset of plant growth and shorten the winter period of food shortage (Post and Stenseth 1999). On the other hand, warmer summers may reduce forage quality as a result of the faster growth to maturity of plants (Bo and Hjeljord 1991, Finstad et al. 2000). Increased precipitation during the summer months can sometimes be favorable for both the amount of forage produced and its nutritional quality, but if excessive, can result in reduced plant growth. Accordingly, contrasting patterns may emerge in these weather influences on herbivore population growth. In Scotland, the responses of red deer to weather variation differed between the west and east coasts (Albon and Clutton-Brock 1988). In central and eastern Scotland, where heather formed the main food source in winter, survival decreased in milder winters which were associated with heavier and hence more prolonged snow cover. On the west coast where the Rum study was conducted, snow cover was not a factor and survival rates were reduced during cold, wet winters. More rainfall during summer negatively affected survival through depressing the growth of grasses and forbs. In contrast, the population growth of elk in Yellowstone National Park was positively related to summer precipitation (Coughenour and Singer 1996), while in Rocky Mountain National Park, elk responded positively to summer rainfall but negatively to high minimum temperatures in summer (Wang et al. 2002). Increasing summer temperatures had a negative influence on the population growth of moose in Minnesota, perhaps by disrupting their thermoregulation (Murray et al. 2006). These authors suggested that continuing warming conditions could lead to a northwards contraction in moose distribution.

In general, conditions during spring and summer affect mainly the survival of juveniles, both in the summer following their birth and over the subsequent winter (Table 3.1). Weather conditions preceding birth affect neonate survival via the nutritional status of their mother (Lubow and Smith 2004). Such influences operate largely through the birth mass and subsequent growth of the offspring, and have enduring effects on the survival and fecundity of the individuals born during favorable years (Albon et al. 1987, Reimers 1995). Both warmer and wetter springs improved the survival of bighorn sheep in Canada (Portier et al. 1998). In northern Canada, the survival of caribou calves was reduced by cool summers (Arthur et al. 2003). Roe deer fawns in France survived better in years with higher rainfall during spring and early summer (Gaillard et al. 1997). The autumn body mass of reindeer calves, and hence their

Table 3.1 Documented effects of temperature and precipitation during spring or summer on population performance of ungulates inhabiting northern temperate regions

Species	Place	Response	Sign	Temperature	Precipitation	Reference
Alces alces	Sweden	Fecundity	+	X		Sand (1996)
Alces alces	Norway	Subsequent recruitment	+	X		Solberg et al. (1999)
Alces alces	Minnesota	Population growth	−	X		Murray et al. (2006)
Antilocapra americana	Texas	Juvenile survival	+		X	Simpson et al. (2007)
Capreolus capreolus	France	Juvenile survival	+		X	Gaillard et al. (1997)
Cervus elaphus	Rum	Juvenile survival	+	X		Albon et al. (1987)
Cervus elaphus	Norway	First reproduction	−	X		Langvatn et al. (1996)
Cervus elaphus	Rocky Mountains	Juvenile survival	+	X	X	Wang et al. (2002)
Cervus elaphus	Jackson	Juvenile survival	+		X	Lubow and Smith (2004)
Cervus elaphus	Yellowstone	Juvenile survival	+		X	Coughenour and Singer (1996)
		Adult survival	+		X	
Odocoileus hemionus	Oregon	Population trend	+		X	Peek et al. (2002)
Ovis canadensis	Alberta	Juvenile survival	+	X	X	Portier et al. (1998)
Rangifer tarandus	Seward Peninsula	First reproduction	+	X	X	Finstad et al. (2000)

See index for common names of species.

"X" indicates presence of an influence from this factor, "sign" whether the influence is positive or negative.

survival prospects, were influenced more strongly by spring conditions controlling the onset of vegetation growth than by snow depth during winter while the offspring were in utero (Pettorelli et al 2005a).

Most mortality beyond the neonatal stage occurs during winter, and is affected by temperature or snow conditions during this period (Table 3.2). Wet and windy weather in March at the end of winter increased mortality among Soay sheep on Hirta (Clutton-Brock et al. 1991, Milner et al. 1999). For reindeer, severe winter weather affected two consecutive cohorts, first causing high calf mortality, then low fecundity among females in the following year (Solberg et al. 2001). For moose, summer temperature as well as snow depth in winter had delayed effects on recruitment through affecting the body condition of females prior to conception (Solberg et al. 1999). For red deer on Rum, temperature conditions and the duration of snow cover in winter affected the proportion of females first giving birth at 3 years of age (Langvatn et al. 1996). In Minnesota, the cumulative effect of snow depth over the preceding three winters on recruitment of moose and white-tailed deer was related partly to the vulnerability of offspring to predation (Mech et al. 1987). Cumulative effects of greater snowfall over consecutive years on population growth were also evident for white-tailed deer in Nova Scotia (Patterson and Power 2002). Deep snow conditions increased mortality especially among juveniles and older elk in Yellowstone (Garrott et al. 2003).Warmer or more variable temperature conditions during early winter can lead to the formation of ice crusts, blocking access to the buried forage (Coughenour et al. 1996, Aanes et al. 2000). An extreme blizzard combining cold temperatures, wind, and heavy snowfall resulted in 50% mortality among pronghorn in Alberta (Barrette 1982).

Variability in terrain and hence in local plant phenology may result in more complex local effects on the growth and survival of herbivores (Mysterud et al. 2001). The timing of the spring onset of vegetation growth is later at higher altitudes, so that ungulates can obtain better quality forage over a more prolonged period by shifting upslope or by exploiting the elevation mosaic in more complex terrain (Albon and Langvatn 1992, Pettorelli et al. 2005b, 2007). Slower snowmelt also extended the period over which high quality vegetation remained available during spring. This landscape variation in plant phenology had beneficial effects on the growth of young red deer and bighorn sheep. However, a more rapid green-up of vegetation during spring had a negative effect on the survival of juvenile bighorn sheep and Alpine ibex, by shortening the period over which high quality forage was available (Pettorelli et al. 2007).

Vulnerability to cold weather or deep snow may be heightened after high density levels have been attained, as shown for bighorn sheep by

Table 3.2 Documented effects of winter weather conditions on population performance of ungulates inhabiting northern temperate regions

Species	Place	Response	Sign	Temperature	Precipitation	Ice crust	Reference
Alces alces	Norway	Subsequent recruitment	+		X		Solberg et al. (1999)
Alces alces	Minnesota	Juvenile survival	−		X		Mech et al. (1987)
		Fecundity	−		X		
Antilocapra americana	Texas	Population trend	+		X		Simpson et al. (2007)
		Juvenile survival					
Bison bison	Yellowstone	Juvenile survival	−		X		Fuller et al. (2007)
Capra ibex	Gran Paradiso	Juvenile survival	−		X		Jacobson et al. (2004)
Cervus elaphus	Yellowstone	Juvenile survival	−		X	X	Coughenour et al. (1996)
		Old survival	−			X	Garrott et al. (2003)
Cervus elaphus	Rum	Juvenile survival	−	X	X		Clutton-Brock et al. (1982)
		Fecundity	+	X	X		Coulson et al. (2000)
Cervus elaphus	Norway	Juvenile survival	+	X			Loison and Langvatn (1998)
Cervus elaphus	Jackson	Survival	+	X			Sauer and Boyce (1983)
Cervus elaphus	Western Scotland	Survival	+	X			Albon and Clutton-Brock (1988)
Cervus elaphus	Eastern Scotland	Survival	−		X		Albon and Clutton-Brock (1988)
Cervus elaphus	Rocky Mountains	Recruitment	+	X			Lubow et al. (2002) Wang et al. (2002)

(continued)

Table 3.2 (*Continued*)

Species	Place	Response	Sign	Temperature	Precipitation	Ice crust	Reference
Odocoileus hemionus	California	Population trend	+		X		Marshal et al. (2002)
Odocoileus virginianus	Nova Scotia	Population trend	−		X		Patterson and Power (2002)
		First reproduction	−		X		
Ovibos moschatus	Norway	Population growth	−		X		Asbjornsen et al. (2005)
Ovis aries	Hirta	Juvenile survival	−		X		Coulson et al. (2001)
		Adult survival	−		X		
Rangifer tarandus	Seward Peninsula	First reproduction	−		X		Finstad et al. (2000)
Rangifer tarandus	Norway	Juvenile survival	+	X			Skogland (1985)
		Juvenile survival	−		X		Weladji and Holand (2006)
Rangifer tarandus	Svalbard	Population trend	−		X	X	Aanes et al. (2000)
							Solberg et al. (2001)
							Chan et al. (2005)
Rangifer tarandus	Svalbard	Population trend and survival	+	X	X	X	Tyler et al. (2008)
Rangifer tarandus	Alaska	Fecundity	−		X		Adams and Dale (1998)
Rangifer tarandus	Finland	Juvenile survival	−	X	X	X	Helle and Kojola (2008)
Rupicapra rupicapra	Pyrenees	Survival	−		X		Gonzalez and Crample (2001)

See index for common names of species.

"X" indicates presence of an influence from this factor, "sign" whether the influence is positive or negative.

Portier et al. (1998). The best model explaining the dynamics of ibex in Gran Paradiso Park incorporated a threshold effect of snow depth relative to population density, rather than the separate effects of these two factors (Jacobson et al. 2004). Lima and Berryman (2006) suggested that deeper snow interacting with nonlinear density dependence shifted the effective carrying capacity for the ibex. However, although the years 1998–2000 had some of the lowest snow depths on record, the ibex population failed to respond to the favorable conditions, indicating that some unidentified factor was also influential.

For Soay sheep on Hirta, the effect of adverse weather conditions depended on the abundance level, the prevailing population structure (Coulson et al. 2001, 2008), and when such weather occurred during the winter months (Hallett et al. 2004). An early pulse of mortality increased survival when bad weather was experienced later, through alleviating the pressure on food resources. After a population crash, the surviving animals consisted largely of prime-aged adult females, the demographic segment most resistant to adverse conditions, promoting rapid population recovery.

To represent the interdependence of the weather effect on the population density, Grenfell et al. (1998) formulated a "self-exciting" threshold autoregressive (SETAR) model. This entailed fitting two linear segments to the density relationship. Two weather variables were incorporated into the model: (i) prevalence of gale-force winds in March in terms of the number of hours spanned, predisposing mortality, and (ii) temperature levels during April, promoting early grass growth and hence rapid population recovery:

$$X_{t+1} = \beta_0 + \beta_1 X_t + c_0 g_{t+1} + d_0 h_{t+1} + \varepsilon \qquad \text{if } X_t \leq X_C$$
$$X_{t+1} = \beta_0 + c_0 g_{t+1} + \varepsilon \qquad \text{if } X_t \geq X_C \qquad (3.2)$$

where $X_t = \log_e(N_t)$, X_C represents the breakpoint density, g indexes gales, and h temperature. This model was modified by Coulson et al. (2001) to represent the contrasting vulnerability of different population segments to extreme weather. The age-structured model incorporating the threshold effect of severe weather conditions explained 90% of the observed variation in annual population growth.

An additional climatic effect important for reindeer and caribou arises through the costs associated with harassment by parasitic insects, which is greater during warmer summers (Klein 1991, Weladji et al. 2002). This affects habitat occupation and time spent foraging, resulting in lowered body mass (Colman 2000) and hence reduced survival prospects.

Rainfall influences on population dynamics have been noted in drier regions at lower latitudes in North America for mule deer in California

(Marshal et al. 2002) and pronghorn in Texas (Simpson et al. 2007). In Arizona and Mexico, the population trend of desert bighorn sheep was positively correlated with winter rainfall and negatively affected by droughts (McKinney et al. 2006, Colchero et al. 2009).

3.1.2 Weather packages

In high northern latitudes, the North Atlantic Oscillation (NAO), indexed by the pressure difference between the Azores and Iceland, is the climatic index mostly widely invoked (Stenseth and Mysterud 2005). A positive NAO value is associated with strong winds bringing warm conditions and elevated precipitation to north-western Europe, while eastern Canada and Greenland experience reduced precipitation. These effects are strongest during the winter months (November–April), and hence also the consequences for ungulate population dynamics (reviewed by Ottersen et al. 2001, Stenseth et al. 2003, Hallett et al. 2004, Stenseth and Mysterud 2005). However, opposite influences may arise even within regions in close proximity, and delayed effects may also be evident.

For example, red deer in Scotland and in maritime regions of Norway responded positively to the NAO index, and hence to warm, wet conditions, while red deer inhabiting interior Norway were adversely affected by such conditions (Forchhammer et al. 1998, Post and Stenseth 1999). This was largely because increased precipitation occurs mostly as rain on the coast, but as deeper snow at higher elevations. On the coast, warmer conditions led to earlier snowmelt and hence earlier onset of plant growth in spring, plus greater spatial variability in time of flowering, lengthening the period over which high quality forage was available.

The time-series model applied by Forchhammer et al. (1998) to various red deer populations took the standard autoregressive form with climatic measures as covariates

$$X_t = \beta_0 + (1 + \beta_1)X_{t-1} + \beta_2 X_{t-2} + \ldots + \beta_d X_{t-d} + \omega_1 W_{t-1}$$
$$+ \omega_2 W_{t-2} + \ldots + \omega_k W_{t-k} + \varepsilon \tag{3.3}$$

where W_t = climate measure at time t, and d and k represent the maximum time lags considered for density and climatic factors respectively. Hence this model incorporated the effects of both direct and delayed climatic effects, indexed by the NAO, as well as direct and delayed density dependence, acting independently. In practice, the lag time d was restricted to a maximum of 3 years for density and k to a maximum of 2 years for climate. The NAO explained 7–19% of the annual change in abundance of these populations, as expressed through the standardized regression coefficients, over and above the effects of density. Direct density dependence exerted its effects mainly through overwinter mortality, while delayed density dependence acted more through effects on growth and fecundity. The lagged positive effect of NAO arose through an increase in the later fecundity of animals born following snowy but warm winters, due to improved forage quality and lengthened period of plant growth in early summer.

Annual changes in the abundance of moose and white-tailed deer in Minnesota were positively correlated with the NAO index 2–3 years previously, i.e. winters that were cold but had a relatively low snow cover promoted subsequent population increase (Post and Stenseth 1998). Overall R^2 values including the additional effects of both ungulate and wolf density were as much as 60% for moose and 87% for the deer, but for moose, the partial correlation with NAO was substantially weaker than for other covariates. The effect of increased snow depth was largely due to increased hunting success of wolves, and the wolf population declined following cold, dry winters.

For the Soay sheep population on Hirta, Stenseth et al. (2004) used the NAO index as the climatic measure, and assumed that extreme weather accentuated the density-dependent decline in the population growth rate above a threshold density (Fig. 3.1a). This "functional coefficient threshold auto-regressive" (FCTAR) model took the form

$$X_t = a_0 + a_1(X_{t-1} - k) + \varepsilon \qquad \text{if } X_{t-1} \leq k$$
$$X_t = a_0 + a_2(X_{t-1} - k) + \varepsilon \qquad \text{if } X_{t-1} \geq k \qquad (3.4)$$

where $k = \log_e(K)$, K represents the zero growth level, $a_0 = \log_e(\lambda) + a_1 k$ where λ is the multiplicative population growth rate, a_1 was set equal to 1 (implying no density effect below the threshold), and $a_2 = a_1 - b$ with b representing a shape parameter. Assuming a simple linear relationship, $a_2 = 1 - b(\text{NAO})$, i.e. b controls the steepness of the population decline above the threshold level, dependent on the NAO conditions.

Including the weather influence improved the coefficient of determination (R^2) of the model fit from 6.0 to 11.4%. Although the fit was less good than obtained from the age-structured model developed by Coulson et al. (2001), far fewer parameters were needed.

After 1995, fluctuations in the Soay sheep population on Hirta shifted towards higher abundance levels than had occurred previously. Berryman and Lima (2006) interpreted this as a "lateral perturbation" (following Royama 1992), with milder weather effectively elevating the carrying capacity. Hence, they fitted the exponential or Ricker version of the logistic

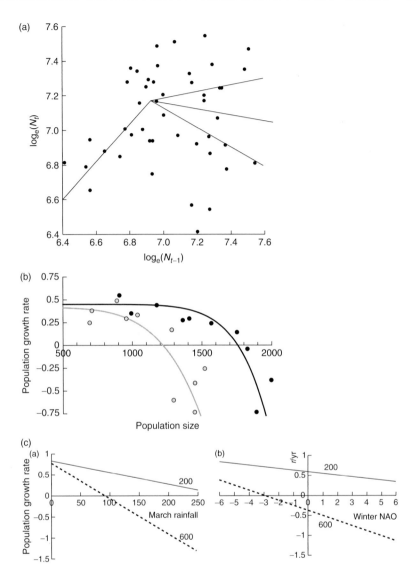

Figure 3.1 Alternative model fitted to the data for Soay sheep dynamics on Hirta. (a) Threshold amplification of density dependence by three levels of winter weather severity, as indexed by the NAO, represented by the threshold autoregressive model of Stenseth et al. (2004), plotting N_{t+1} versus N_t. (b) Theta-logistic model of Berryman and Lima (2006), comparing nonlinear relationship between the population growth rate and population size before 1995 (*grey*) with that from 1995 onward when milder weather conditions prevailed (*black*). (c) Regression relationships for effects on the population growth rate of winter severity represented by March rainfall and for the NAO influence derived by Hone and Clutton-Brock (2007), comparing high (*dashed line*) versus low (*solid line*) abundance levels.

equation incorporating curvilinear density dependence:

$$r_t = r_{max}\{1 - (N_{t-1}/K)^\theta\}$$

with the effective value of K dependent on the prevailing climatic conditions. They justified the convex form of density dependence as representing scramble competition (Fig. 3.1b), and assumed that climatic conditions, indexed by the NAO, influenced the energy demand of the sheep and hence the amount of food needed to survive through winter. An influence of warmer temperature conditions during summer on food production and hence population growth was considered, but inadequately supported statistically. Lowered values for the NAO, and consequently milder winters, underlie the changed conditions between 1995 and 2005.

Hone and Clutton-Brock (2007) assessed the fit of somewhat simpler models to the population dynamics of the Soay sheep, and red deer on Rum. The climatic influence was represented by either March rainfall or winter NAO, acting additively on the population growth rate (Fig. 3.1c). For Soay sheep, the best supported model ($R^2 = 49\%$) represented the weather effect as increasing the weighting of population size in an additional density feedback term

$$(N_{t+1} - N_t)/N_t = r_m\{(1 - aN_t/K) - (bWN_t/K)\} \qquad (3.5)$$

where W represents the climate measure and K indexes food availability. This model replicated the dynamics of red deer nearly as well as for the sheep ($R^2 = 45\%$). However, March rainfall affected red deer positively but Soay sheep negatively, although the effects of NAO were consistently negative for both species. Catchpole et al. (2004) found that the NAO index provided a better fit to the survival rates of both male and female red deer than any of the local weather covariates.

A further model comparison for Soay sheep by Coulson et al. (2008) partitioned population growth into separate terms for recruitment and survival:

$$\lambda = (1 + \varepsilon L)/\{1 + \exp^{-(a + bN + cW + dNW + ...)}\} \qquad (3.6)$$

where L = average litter size, ε is a modifying constant, and the weather influence W is represented by the NAO. This model explained almost 70% of the variation in Soay sheep abundance between 1985 and 2001, with the NAO influence contributing 55%. A fully age-structured model showed a lesser influence of the NAO (22%), because the weather effect operated largely through differential survival among the demographic classes. The combined effect of density and weather yielded a linear fit between predicted

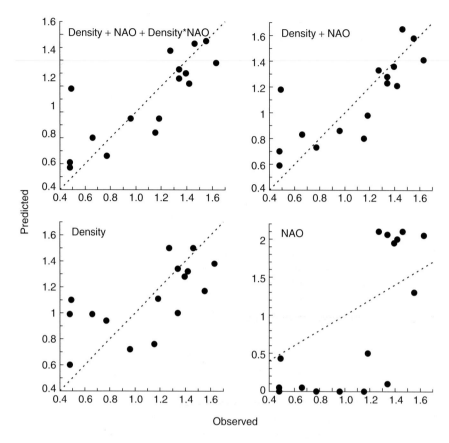

Figure 3.2 Relationships between observed and predicted growth rates of the Soay sheep population, expressed as N_{t+1}/N_t, projected by the interactive model developed by Coulson et al. (2008), comparing full model including density, weather, and their interaction (*top left*), additive effects of density and weather alone (*top right*), density alone (*bottom left*), and weather alone (*bottom right*).

and observed values, whereas the density effect alone showed a curvilinear relationship and the weather effect took the form of separate clusters of high versus low deviations (Fig. 3.2).

Common weather influences can lead to regional synchrony in the population dynamics of different ungulate species (Grotan et al. 2005). Caribou and muskox showed opposite patterns on the eastern and western coasts of Greenland, due to contrasting effects of NAO on temperature conditions during winter in these regions (Post and Forchhammer 2002). Furthermore, muskox abundance tended to decline following warm and hence snowy winters, while caribou were adversely affected by cold dry winters (Forchhammer et al. 2002). Climate as indexed by the NAO explained up to 24% of the variation in annual population growth for muskox and up to 16% for caribou.

The Arctic Oscillation (AO), indexed by atmospheric pressure contrasts across the arctic region, explained the population dynamics of reindeer in Svalbard ($78°N$) somewhat better than the NAO (Aanes et al. 2002). A high value is associated with cloudy, cool summers reducing plant growth, and deeper snow during winter, leading to lowered population growth by the reindeer. In one year, extreme ice crust formation caused a severe crash by the reindeer population, and Chan et al. (2005) suggest ways to model the effects of such infrequent events. Contrary to expectations, an "ablation" index, representing the frequency and magnitude of thaw-freeze cycles contributing to the hardness of the snowpack, showed a positive association with overwinter survival and hence population trend of reindeer in another section of Svalbard (Tyler et al. 2008). This was because warm conditions in early winter commonly caused most snow cover to melt, enhancing forage availability in late winter. Prevalently low values of NAO in recent years have promoted an increasing trend by these reindeer. Semi-domesticated reindeer in Finnish Lapland, where the snow cover is typically deep, showed higher overwinter mortality during winters with positive values of both the NAO and AO, associated with warmer temperatures and more frequent icing of the snow cover (Helle and Kojola 2008). Snow depth as well as icing conditions also substantially reduced calf survival over winter. For this population, NAO conditions alone explained 18% of the overwinter variation in mortality. In Banff National Park in western Canada, the North Pacific Oscillation influenced the population dynamics of elk through its effects on snow depth and hence the vulnerability of elk to predation by wolves (Hebblewhite 2005).

3.2   Tropical and subtropical environments

3.2.1   Individual weather components

Rainfall is seasonally restricted through much of the tropics and subtropics, excluding the equatorial forest zone. Near the equator, two wet seasons follow the passage of the intertropical convergence in atmospheric circulation. Towards subtropical latitudes, there is a single rainy season during the summer months. Further south or north, Mediterranean-type conditions prevail with dry summers and precipitation mostly in winter. During the dry season, grasses become dormant and retain little green foliage towards the end. Woody plants are predominantly deciduous and shed their leaves, with the remaining evergreen foliage highly fibrous and mostly chemically defended. Hence, large herbivores experience a bottleneck in food quality and quantity towards the end of the dry season.

Because of the erratic nature of precipitation, the amount of rainfall received annually varies quite widely from year to year, with consequent effects on the vegetation growth supporting herbivores. To the extent that the amount of food produced determines the abundance of herbivores, rainfall variation can be interpreted as modifying the effective carrying capacity K in the logistic model. Taking into account also the effect of the abundance level in the availability of this food per capita, the population growth rate can be related inversely to the effective food share expressed through the rainfall : abundance ratio:

$$N_{t+1} - N_t = r_0 N_t (1 - N_t / R_t) \qquad (3.7)$$

where N_t represents some measure of herbivore abundance, and R_t represents a measure of rainfall. Alternatively, the survival rates contributing to the change in abundance can be expressed as directly dependent on rainfall divided by the population density:

$$S_{a,t+1} = f(R_t / N_t) \qquad (3.8)$$

where $S_{a,t+1}$ = survival rate into stage class a between times t and $t+1$. Instead, the annual mortality rate could be expressed as a function of the population density relative to the annual rainfall. For kudu, this functional relationship explained over 85% of the annual variation in juvenile survival, and around 50% of survival variation among adult females (Owen-Smith 1990, 2006; Fig. 3.3).

The stage-specific survival functions were used to model the dynamics of the kudu population in response to rainfall variation (Owen-Smith 2000). The rainfall measure was the annual total, while the population abundance was expressed as biomass density to take account of the varying age structure. A log-linear relationship was initially chosen for the survival relationship, assuming that the effect of rainfall varied proportionately rather than through increments in amount. This model closely replicated the observed dynamics of the kudu population over the study period (Fig. 3.3d).

The population dynamics of buffalo and waterbuck, likewise, appeared strongly positively dependent on annual rainfall variation in the Kruger National Park (Mills et al. 1995). However, other ungulate species appeared less responsive to rainfall, while wildebeest and zebra tended to increase during periods when the annual rainfall remained generally low. Smuts (1978) suggested that the vulnerability of wildebeest and zebra to predation

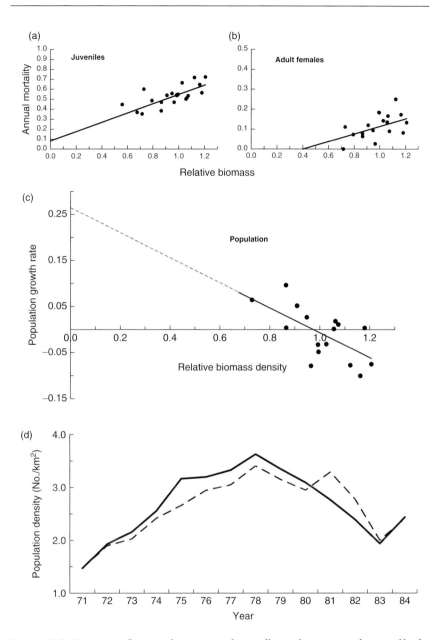

Figure 3.3 Stage-specific mortality rates and overall population growth rate of kudu related to the population biomass density relative to the annual rainfall total, both normalized around their mean values. (a) Juvenile mortality loss derived from the cow : calf ratio. (b) Mortality rate of prime and old adults combined. (c) Overall population growth rate as a proportion. (d) Modeled population trend projected from these relationships (*dashed line*), in comparison with the observed population growth rate (*solid line*) (from Owen-Smith (2000, 2006)).

in Kruger Park was greater during years with high rainfall through the effect of grass growth on cover for stalking lions. The population dynamics of wildebeest in both Kruger and Serengeti appeared to be related to the amount of rainfall received during the normally dry season months, influencing the amount of green leaf remaining through this critical period (Mduma et al. 1999, Ogutu and Owen-Smith 2003). In contrast to the situation in Kruger Park, the population growth of zebra in the Laikipia region of Kenya was positively related to the annual rainfall, largely through the high mortality incurred during drought years (Georgiadis et al. 2003). Notably, zebra were less susceptible to predation at Laikipia than in Kruger Park.

Ogutu and Owen-Smith (2003) evaluated the separate effects of seasonal rainfall components, as well as the influence of temperature conditions potentially affecting grass growth and green leaf retention, on the population dynamics of various large ungulates in Kruger Park over a period when many species showed substantial population declines. They fitted a standard autoregressive model structure to the time series of annual census totals for these species, augmented by additional predictors for the weather effects:

$$X_t = \beta_0 + (1 + \beta_1)X_{t-1} + \beta_2 R_W + \beta_3 R_D + \beta_4 T + \varepsilon \qquad (3.9)$$

where X_t = count total (log-transformed), R_W = wet season rainfall (log-transformed), R_D = preceding dry season rainfall (log-transformed), and T = deviation in mean daily temperature from the long-term mean. Dry season rainfall had a significant influence on the dynamics of seven out of eleven species, while five species responded significantly to the wet season component. Wildebeest appeared uninfluenced by variation in wet season rainfall, while zebra showed only a weak response to both seasonal components. Giraffe appeared quite unaffected by rainfall in any season. Temperature had no additional influence on the dynamics of any species.

Besides the direct effect of rainfall on plant growth and green leaf retention, changes in rainfall can also affect the grass cover and composition, such that the herbaceous layer could become less productive for the same rainfall. Accordingly, Owen-Smith and Mills (2006) considered additionally the lagged effect of prior rainfall conditions on the dynamics of these same ungulate populations:

$$r_t = \beta_0 + \beta_A A_t + \beta_R R_t + \beta_3 R_D + \beta_4 R_{t-\tau} + \varepsilon \qquad (3.10)$$

where $r_t = \log_e(N_{t+1}/N_t)$, A_t is some measure of population abundance, R_t is some measure of current rainfall, and $R_{t-\tau}$ represents the lagged influence of rainfall cumulative over a period τ back into the past. Models incorporating the aggregate effect of annual rainfall over the preceding 5 years in addition to current effects of the seasonal rainfall components were most strongly supported by the data, as judged by Akaike information criterion (AIC), for seven of these eleven ungulate populations. However, the lagged effect could also have arisen through changes in predator abundance in response to changing prey availability, because the period with low prior rainfall was associated with high past prey abundance for predators plus low current dry season rainfall. The additional influence of past conditions, whether through predation or habitat deterioration, was evident from the lowered population growth relative to rainfall shown by the declining populations after 1986 (Fig. 3.4).

In the Mara-Serengeti region with distinct early and late wet seasons, the abundance of seven ungulate species was curvilinearly related to both the current rainfall and cumulative past rainfall (Ogutu et al. 2008a). Zebra were anomalous in showing a negative rather than positive relationship with immediately prior rainfall. For the more mobile species, changes in numbers counted probably reflect movements in and out of the study region rather than intrinsic changes in populations, particularly for the migratory zebra that concentrate in this region during the dry season. For topi, water-buck, and warthog, juvenile proportions were closely related to rainfall over

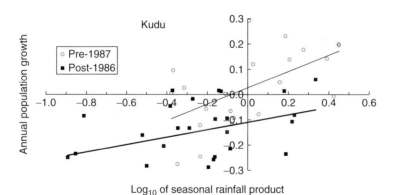

Figure 3.4 Relationship between annual population growth and rainfall for kudu in Kruger Park, as affected by a change in predation pressure. The period after 1986, associated with elevated predation (*closed squares*), is compared with earlier years (*open circles*). Rainfall is represented by the product of the wet and dry season components normalized relative to their means, then log-transformed (from Owen-Smith and Mills 2006).

the 6 months preceding their birth. However, for impala, hartebeest, and perhaps giraffe, the number of immature animals present was most closely associated with late-wet season rainfall over the preceding 2–5 years.

In severe drought years, most large herbivore species show elevated mortality and consequent population declines arising through food shortages by the late dry season. In the Serengeti region, severe drought conditions in 1993 caused 70% mortality among buffalo and 40% mortality among wildebeest (Sinclair et al. 2007). Although wildebeest and zebra populations were unaffected by the El Nino-related drought of 1982/3 in Kruger Park, in the nearby Klaserie Private Nature Reserve these species declined by 80% or more of their pre-drought abundance (Walker et al. 1987). The crucial difference in conditions lay in the availability of surface water. Numerous water points in the private reserve elevated population abundance in the short term, but led to more widespread depletion of forage when drought conditions took hold. During the same drought period, wildebeest crashed to 12% of their former numbers in the Central Kalahari region of Botswana, with a fence blocking migration towards remaining sources of surface water being a contributory factor (Spinage and Matlhare 1992). Hartebeest, eland, and kudu showed population declines by 50% or more, while the springbok population was less affected. Animals surviving droughts may show reduced calf production in the following year (Estes et al. 2006). While the adverse effects of drought years on large herbivore population have been widely reported in Africa, fewer studies have related population growth to less extreme variation in rainfall (Table 3.3).

Rainfall also strongly influenced the dynamics of kangaroos in semi-arid regions of Australia, where rainfall patterns are highly erratic without a strong seasonal component (Bayliss 1985). The 1982/3 drought was associated with population declines amounting to 40% on average across a vast region, with most deaths occurring during the summer months after the expected rains had failed (Caughley et al. 1985).

Because animals in tropical environments must cope with midday temperatures commonly exceeding 35°C, they may be vulnerable to hypothermia at temperature levels that would not affect northern ungulates. Kudus showed elevated mortality in a year when exceptionally cold and wet conditions occurred in the late dry season (Owen-Smith 1990, 2000). Extreme conditions were indicated by maximum daily temperature not rising above 14°C, rather than by unusually low minimum temperatures, with heat loss accentuated by rain and wind at a stage when body fat reserves would be near their minimum. Cold-related die-offs have also been recorded for reedbuck in Zimbabwe (Ferrar and Kerr 1971) and nyala in South Africa (Keep 1973). However, other species in the same region were affected little or not at all by the same weather conditions.

Table 3.3 Documented effects of rainfall (annual or seasonal components), or only extreme drought conditions, on performance measures for ungulate populations in tropical or subtropical latitudes

Species	Place	Response	Annual	Wet season	Dry season	Drought	Reference
Aepyceros melampus	Klaserie PNR	Abundance				X	Walker et al. (1987)
Aepyceros melampus	Kruger NP	Juvenile survival Population trend		X	X		Owen-Smith et al. (2005) Ogutu and Owen-Smith (2003)
Alcelapus buselaphus	Central Kalahari	Abundance				X	Spinage and Matlhare (1992)
Alcelapus buselaphus	Nairobi NP	Abundance				X	Hillman and Hillman (1977)
Antidorcas marsupialis	Central Kalahari	Abundance				X	Spinage and Matlhare (1992)
Connochaetes taurinus	Serengeti NP	Juvenile survival			X		Mduma et al. (1999)
Connochaetes taurinus	Kruger NP	Juvenile survival Adult survival			X X		Owen-Smith et al. (2005)
Connochaetes taurinus	Klaserie Private Reserve	Abundance				X	Walker et al. (1987)
Connochaetes taurinus	Central Kalahari	Abundance				X	Spinage and Matlhare (1992)
Connochaetes taurinus	Ngorongoro	Juvenile survival				X	Estes et al. (2006)
Damaliscus dorcas	Bontebok NP	Recruitment	X				Novellie (1986)

(continued)

Table 3.3 (Continued)

Species	Place	Response	Annual	Wet season	Dry season	Drought	Reference
Damaliscus lunatus	Kruger	Juvenile survival			X		Owen-Smith et al. (2005)
Damaliscus lunatus	Zimbabwe ranch	Adult survival		X			Dunham et al. (2003)
		Juvenile survival			X		
Equus burchelli	Klaserie PNR	Abundance				X	Walker et al. (1987)
Equus burchelli	Kruger	Population trend		X	X	X	Ogutu and Owen-Smith (2003)
Equus burchelli	Laikipia Ranches	Population trend	X				Georgiadis et al. (2003)
Giraffa camelopardalis	Klaserie PNR	Abundance				X	Walker et al. (1987)
Giraffa camelopardalis	Kruger NP	Juvenile survival			X		Owen-Smith et al. (2005)
Hippotragus equinus	Kruger NP	Juvenile survival		X			Owen-Smith et al. (2005)
Hippotragus niger	Kruger NP	Adult survival			X		Owen-Smith et al. (2005)
Kobus ellipsiprymnus	Klaserie PNR	Abundance				X	Walker et al. (1987)
Kobus ellipsiprymnus	Kruger NP	Juvenile survival			X		Owen-Smith et al. (2005)
		Adult survival			X		
		Population trend		X	X		Ogutu and Owen-Smith (2003)
Kobus kob	Sudan	Adult survival		X			Fryxell (1987)
Oryx gazella	Kalahari	Abundance				X	Spinage and Matlhare (1992)

Species	Location	Parameter					Reference
Phacochoerus aethiopicus	Klaserie PNR	Abundance	X				Walker et al. (1987)
Phacochoerus aethiopicus	Kruger NP	Population trend		X	X		Ogutu and Owen-Smith (2003)
Syncerus caffer	Klaserie PNR	Abundance	X				Walker et al. (1987)
Syncerus caffer	Kruger NP	Population trend				X	Mills et al. (1995)
Syncerus caffer	Masai Mara NR	Abundance	X				Sinclair et al. (2007)
Syncerus caffer	Ngorongoro	Juvenile survival	X				Estes et al. (2006)
		Adult survival					
Taurotragus oryx	Kruger NP	Population trend		X			Ogutu and Owen-Smith (2003)
Taurotragus oryx	Central Kalahari	Abundance	X				Spinage and Matlhare (1992)
Taurotragus oryx	Southern Kalahari	Abundance	X				Knight (1995)
Tragelaphus strepsiceros	Klaserie PNR	Abundance	X				Walker et al. (1987)
Tragelaphus strepsiceros	Kruger NP	Juvenile survival		X	X		Owen-Smith et al. (2005)
		Adult survival		X	X		
Tragelaphus strepsiceros	Central Kalahari	Abundance	X				Spinage and Matlhare (1992)
Macropus spp	New South Wales	Population trend				X	Bayliss (1985)
Macropus rufus	South Australia	Population trend				X	Cairns and Griggs (1993)

See index for common names of species.

"X" indicates presence of an influence from this factor, "sign" whether the influence is positive or negative.

3.2.2 Weather packages

Rainfall variation through much of the subtropics and parts of the tropics is strongly influenced by the El Nino–Southern Oscillation (ENSO), governed by sea surface temperature in the eastern Pacific Ocean and indexed by the atmospheric pressure difference between Tahiti and Darwin (the Southern Oscillation Index or SOI; McPhaden et al. 2006). Warm temperature conditions (the "El Nino" phase) are associated with prevalently high pressure over eastern southern Africa and eastern Australia around the southern summer solstice, suppressing midsummer rainfall. At the same time, eastern equatorial Africa and California typically experience elevated rainfall (Tyson and Gatebe 2001). The opposite "La Nina" phase occurs when cool ocean temperatures prevail in the eastern Pacific. In tropical East Africa, ENSO has little influence on the long rains through March–April (Ogutu et al. 2008b). Dry season rainfall received during the southern winter also varies independently of ENSO. Furthermore, a quasi-decadal oscillation in rainfall has prevailed in the summer rainfall region of South Africa over the past century, not obviously associated with ENSO (Tyson 1991). Additional influences on African rainfall come from the Indian Ocean dipole, which may modify or counteract the influence from the Pacific Ocean (Marchant et al. 2007).

However, the SOI explained substantially less of the variation in ungulate population dynamics than direct rainfall measures, except for species that responded little to rainfall variation (Ogutu and Owen-Smith 2003, Marshal et al. unpublished). For African buffalo, the best model incorporating SOI accounted for 39% of annual variation in population growth, compared with 60% for the best rainfall-related model, including the separate contributions of both the wet and dry season components. The ENSO influence was apparent mainly in the extreme drought years of 1982/3 and 1991–2. Furthermore, the severe El Nino conditions experienced worldwide during 1997/8 had little effect on southern African rainfall, perhaps being counteracted by the Indian Ocean dipole.

3.3 Effects of predation and hunting

The risk of predation may restrict foraging to more secure habitats, increasing pressure on resources in these habitats (Fortin et al. 2005). Through this mechanism, populations may become food-limited at abundance levels somewhat lower than could be supported by resources if animals foraged more widely. Foraging time may also be restricted by time diverted to vigilance (Laundre et al. 2001). On the other hand, food deficiencies could prompt animals to extend their foraging into more risky habitats, thereby amplifying mortality among prime-aged animals that would not have died from the nutritional shortfalls alone (Sinclair and Arcese 1995). In the absence of hunting or predation, populations may show greater responses to climatic influences, because older animals more susceptible to mortality make up a greater proportion of the adult segment.

Wang et al. (2009) found that density dependence was weakened in northern ungulate populations in the presence of large carnivores, as measured by the coefficient β_1 in an autoregressive model (see eqn. (3.9) above). No evidence for density dependence was found in ungulate populations in very high northern latitudes, either because variable weather obscured the density feedback, or because the prevalence of large carnivores in these regions attenuated the density influence. Nevertheless, density dependence was evident for four out of thirteen ungulate populations occupying subtropical latitudes in Kruger Park, after controlling for rainfall variation, and appeared negative in most cases even when statistical support was weak, despite the abundance of large carnivores (Ogutu and Owen-Smith 2003). A subsequent analysis revealed density effects on the population growth and survival also of kudu and wildebeest in this area, despite the high susceptibility of these two species to predation (Owen-Smith 2006). Effects of rainfall on vulnerability to predation seemed mainly responsible for fluctuations in abundance of various ungulate populations in Kruger Park (Ogutu and Owen-Smith 2005). Shifting prey selection related to the changing vulnerability of the principal prey species, dependent on rainfall as well as population status, led to severe population declines by less common ungulate species (Owen-Smith and Mills 2008).

Wilmers et al. (2007) suggested that largely additive predation by ambush predators such as lions could dampen fluctuations in prey abundance, by restricting the rate of population recovery after weather-induced declines. Cursorial predators capturing mainly weakened prey could stabilize prey dynamics, through amplifying mortality among old and young animals and thereby reducing pressure on food resources. The population impact of wolves on the abundance of elk in northern Yellowstone seemed to be much less than that of human hunters (Vucetich et al. 2005), because the mean age of elk in hunter kills was 6.5 years, compared with more than 12 years for elk killed by wolves (Wright et al. 2006). Nevertheless, repeated disturbance of elk activity by the wolves and restrictions on habitat use mean that the presence of this predator potentially has a much greater impact on the elk population than simply through mortality imposed (White and Garrott 2005, Creel et al. 2007). Following the establishment of the wolf population, deaths of elk calves ascribed to malnutrition during winter decreased drastically to zero (Barber-Meyer et al. 2008).

3.4 Overall assessment

Variations in precipitation have a pervasive influence on the population dynamics of large herbivores, largely through effects on food production and its seasonal availability. The consequences are evident through changes in the birth mass and growth rate of animals, and hence reproductive success as well as vulnerability to mortality. Because rainfall controls plant growth as well as seasonal retention of green foliage through the dry season in tropical savannas, there is a positive association between increased rainfall and population growth for perhaps the majority of ungulate species inhabiting such ecosystems. Nevertheless, populations of some species seemed unresponsive to the consequent variation in food production (e.g. zebra and giraffe in Kruger Park), or affected only by the dry season component (wildebeest in Kruger Park and Serengeti). However, severe drought conditions greatly reducing the amount of forage persisting through the dry season lead to elevated mortality among almost all herbivore species.

In higher latitudes, increased summer rainfall in regions experiencing relatively dry summers likewise promotes improved offspring survival. However, the effect of greater rainfall becomes negative in regions where summers are generally wet. Greater precipitation in the form of snow during winter is also adverse for most northern ungulates, through decreasing forage accessibility and increasing the costs of movement. In all circumstances where weather conditions affect food availability, the consequences for population dynamics depend also on population density, determining the demand for the limiting resources.

The effects of temperature variation are less consistent. Generally, warmer winters tend to be associated with greater snowfall in north temperate latitudes, and also more frequent thaw-freeze cycles generating ice crusts that hamper foraging. The latter conditions can lead to severe mortality among northern ungulates, but are not revealed by mean temperature measures. In regions experiencing little snowfall, wet and windy conditions seem more influential than extreme cold in inducing mortality from cold stress. Even subtropical ungulates may be subject to elevated mortality when such conditions occur towards the end of the dry season. Elevated temperatures during summer appeared adverse for moose.

Susceptibility to predation may mediate the effects of weather conditions in some circumstances. Deeper snow hampers the ability of northern ungulates to escape from wolves, while greater rainfall results in taller grass, and hence better concealment for stalking predators like lions.

In temperate latitudes, the aggregate measure of associated weather conditions represented by the NAO seemed usually to be superior to specific measures of local weather conditions in predicting herbivore population trends. However, the expression of the NAO on local weather is spatially disparate, so that local ungulate populations may experience contrasting conditions in the same year. I was surprised by how little of the annual variation in population growth was explained by NAO for any northern ungulate population besides Soay sheep. Although applications of the SOI to tropical or subtropical ungulate populations are few, its predictive power seemed to be somewhat better, despite being outperformed by direct rainfall measures. The effects of ENSO are more widespread than just in the tropics, and weather patterns associated with NAO are more pervasive than just in northern temperate latitudes, but no study has yet considered the combined effects of these atmospheric conditions. Considerations of NAO for northern ungulates have almost invariably been restricted to winter conditions, and little attention has been given to influences on summer weather conditions affecting offspring recruitment.

While the possibility was raised that the presence of predation or human hunting could modify the expression of weather influences on population dynamics, there have been few demonstrations of such effects. For wildebeest and zebra in Kruger Park, shifting vulnerability to predation apparently counteracted whatever beneficial effects the amount of rainfall received might have had on food production. For northern ungulates, the expression of density dependence was weakened in the presence of large carnivores, but weather influences were not considered.

Assessments have generally been restricted to establishing the relatively immediate effects of prevailing weather conditions on population growth, survival, or recruitment rates, with little attention given to the longer term or delayed effects of past weather on habitat conditions affecting resource availability or security from predation. Through affecting fire frequency and intensity, rainfall conditions can alter the balance between woody plants and grasses in African savannas (Dublin et al. 1990). In higher northern latitudes, forest fires exacerbated by hot, dry, and windy weather conditions also have a major impact on the patch mosaic of vegetation structure and composition (Turner et al. 1994). Intense windstorms fell forest trees, generating a diversity in vegetation structure that can be exploited by ungulates. In tropical savannas, higher CO_2 concentrations in the atmosphere can confer a growth advantage on woody plants dependent on the C3 photosynthetic pathway over tropical grasses adapted to handle hot dry conditions through a C4 pathway (Bond and Midgley 2000). Substantial parts of the Serengeti ecosystem are becoming invaded by trees at the expense of open grassland (Sinclair et al. 2007). Greater woody cover not only reduces the amount of grass produced, but also increases the vulnerability of ungulates to predation.

The vast literature summarized in this chapter provides much information on how ungulate populations respond to weather variation, but little information to judge how they might be affected by global climate change. Warmer conditions do not immediately seem to present much threat, apart from more frequent ice crusting in areas with a winter snow cover. Of greater concern are projections that global warming is likely to promote more frequent and intense storms, associated with flooding and windthrow of trees. What might the consequent changes in local habitat conditions portend for large herbivores?

In tropical latitudes, the expectation is that global warming will lead to more variable weather conditions, expressed through more frequent floods and droughts. Hence models need to consider not just the potential response of large herbivore populations to greater or lesser mean rainfall in tropical latitudes, but also the consequences of greater annual variance in rainfall. Convex response functions to diminishing resource availability, whether due to rising density or lesser food production, mean that the consequences of adverse years weigh more heavily on population trend than the compensatory potential in good times. Furthermore, the potential for ungulates to counteract locally adverse conditions through exploiting regional ecosystem heterogeneity is becoming circumscribed as populations become increasingly confined within designated protected areas. The models that are needed must be able to accommodate these larger scale and longer term consequences of global changes in both climate and human domination of the biosphere. The modeling elaborations needed are the subject of the final chapter.

Before confronting them, intervening chapters need to address demographic processes in more detail, the propensity of large herbivore populations to fluctuate in abundance through interactions with vegetation and predators, and the implications of spatial heterogeneity for population processes.

Acknowledgments

Helpful comments on various drafts of this chapter were provided by Nils Stenseth, Jason Marshal, Joris Cromsigt, and John Fryxell.

References

Aanes, R., B.-E. Saether, and N. A. Oritsand. 2000. Fluctuations of an introduced population of Svalbard reindeer: the effects of density dependence and climatic variation. *Ecography* 23: 437–443.

Aanes, R., B.-E. Saether, F. M. Smith, E. J. Cooper, P. A. Wookey, and N. A. Oritsand. 2002. The Arctic Oscillation predicts effects of climate change in two trophic levels in a high-arctic ecosystem. *Ecology Letters* 5: 445–453.

Adams, L. G. and B. W. Dale. 1998. Reproductive performance of female Alaskan caribou. *Journal of Wildlife Management* 62: 1184–1195.

Albon, S. D. and T. H. Clutton-Brock. 1988. Climate and the population dynamics of red deer in Scotland. In *Ecological Changes in the Uplands*, eds. M. B. Usher and D. B. A. Thompson, pp. 93–107. Blackwell, Oxford.

Albon, S. D., T. H. Clutton-Brock, and F. E. Guinness. 1987. Early development and population dynamics in red deer. II. Density-independent effects and cohort variation. *Journal of Animal Ecology* 56: 69–81.

Albon, S. D. and R. Langvatn. 1992. Plant phenology and the benefits of migration in a temperate ungulate. *Oikos* 65: 502–513.

Arthur, S. M., K. R. Whitten, F. J. Mauer, and D. Cooley. 2003. Modeling the decline of the Porcupine Caribou Herd 1989–1998: the importance of survival versus recruitment. *Rangifer* Special Issue No.14: 123–130.

Asbjornsen, E. J., B.-E. Saether, J. D. C. Linnell, S. Engen, R. Andersen, and T. Bretten. 2005. Predicting the growth of a small introduced muskox population using population prediction intervals. *Journal of Animal Ecology* 74: 612–618.

Barber-Meyer, S. M., L. D. Mech, and P. J. White. 2008. Elk Calf Survival and Mortality Following Wolf Restoration to Yellowstone National Park. *Wildlife Monographs* No. 169, 30 pp.

Barrette, M. W. 1982. Distribution, behavior and mortality of pronghorns during a severe winter in Alberta. *Journal of Wildlife Management* 46: 991–1002.

Bayliss, P. 1985. The population dynamics of red and western grey kangaroos in arid New South Wales, Australia. I. Population trends and rainfall. *Journal of Animal Ecology* 54: 111–125.

Berryman, A. and M. Lima. 2006. Deciphering the effects of climate on animal populations: diagnostic analysis provides new interpretation of Soay sheep dynamics. *American Naturalist* 168: 784–795.

Bo, S. and O. Hjeljord. 1991. Do continental moose ranges improve during cloudy summers? *Canadian Journal of Zoology* 69: 1875–1879.

Bond, W. J. and G. F. Midgley. 2000. A proposed CO_2-controlled mechanism of woody plant invasion in grasslands and savannas. *Global Change Biology* 6: 865–870.

Cairns, S. S. and G. C. Grigg. 1993. Population dynamics of red kangaroos in relation to rainfall in the South Australian pastoral zone. *Journal of Applied Ecology* 30: 444–458.

Catchpole, E. A., Y. Fan, B. T. T. Morgain, T. Clutton-Brock, and T. Coulson. 2004. Sexual dimorphism, survival and dispersal in red deer. *Journal of Agricultural, Biological, and Environmental Statistics* 9: 1–26.

Caughley, G., G. C. Grigg, and L. Smith. 1985. The effect of drought on kangaroo populations. *Journal of Wildlife Management* 49: 679–685.

Chan, K.-S., A. Mysterud, N. A. Ortland, T. Severinsen, and N. C. Stenseth. 2005. Continuous and discrete extreme climatic events affecting the dynamics of a high-arctic reindeer population. *Oecologia* 145: 556–563.

Chesson, P. 2003. Understanding the role of environmental variation in population and community dynamics. *Theoretical Population Biology* 64: 253–254.

Clutton-Brock, T. H., F. E. Guinness, and S. D. Albon. 1982. *Red Deer: Behavioural Ecology of Two Sexes*. University of Chicago Press, Chicago.

Clutton-Brock, T. H., O. F. Price, S. D. Albon, and P. A. Jewell. 1991. Persistent instability and population regulation in Soay sheep. *Journal of Animal Ecology* 60: 593–608.

Colchero, F., R. A. Medellin, J. S. Clark, R. Lee, and G. G. Katul. 2009. Predicting population survival under future climate change: density dependence, drought and extraction in an insular bighorn sheep population. *Journal of Animal Ecology* 78: 666–673.

Colman, J. E. 2000. Behaviour Patterns of Wild Reindeer in Relation to Sheep and Parasitic Flies. PhD thesis, University of Oslo, Norway.

Coughenour, M. B. and F. J. Singer. 1996. Elk population processes in Yellowstone National Park under the policy of natural regulation. *Ecological Applications* 6: 573–593.

Coulson, T., E. A. Catchpole, S. D. Albon, et al. 2001. Age, sex, density, winter weather, and population crashes in Soay sheep. *Science* 292: 1528–1531.

Coulson, T., T. H. G. Ezard, F. Pelletier, et al. 2008. Estimating the functional form of the density dependence from life history data. *Ecology* 89: 1661–1674.

Coulson, T., E. J. Milner-Gulland, and T. Clutton-Brock. 2000. The relative roles of density and climatic variation on population dynamics and fecundity rates in three contrasting ungulate species. *Proceedings of the Royal Society of London B: Biological Sciences* 267: 1771–1779.

Creel, S., D. Christianson, S. Liley, and J. A. Winnie. 2007. Predation risk affects reproductive physiology and demography of elk. *Science* 315: 960.

Dublin, H. T., A. R. E. Sinclair, and J. McGlade. 1990. Elephants and fire as causes of multiple stable states in the Serengeti-Mara woodlands. *Journal of Animal Ecology* 59: 1147–1164.

Dunham, K. M., E. F. Robinson, and C. M. Swanepoel. 2003. Population decline of tsessebe antelope on a mixed cattle and wildlife ranch in Zimbabwe. *Biological Conservation* 113: 111–124.

Estes, R. D., J. L. Atwood, and A. B. Estes. 2006. Downward trends in Ngorongoro Crater ungulate populations 1986–2005: conservation concerns and the need for ecological research. *Biological Conservation* 131: 106–120.

Ferrar, A. A. and M. A. Kerr. 1971. A population crash of the reedbuck in Kyle National Park, Rhodesia. *Arnoldia (Rhodesia)* 5 (16): 1–19.

Finstad, G. L., M. Berger, K. Kielland and A. K. Pritchard. 2000. Climatic influence on forage quality, growth and reproduction of reindeer on the Seward Peninsula. II. Reindeer growth and reproduction. *Rangifer* Special Issue No.12: 144.

Forchhammer, M., N. C. Stenseth, E. Post, and R. Langvatn. 1998. Population dynamics of Norwegian red deer: density dependence and climatic variation. *Proceedings of the Royal Society of London B: Biological Sciences* 265: 341–350.

Forchhammer, M., E. Post, N. C. Stenseth, and D. M. Boertmann. 2002. Long-term responses in arctic ungulate dynamics to changes in climate and trophic processes. *Population Ecology* 44: 113–120.

Fortin, D., H. L. Beyer, M. S. Boyce, D. W. Smith, T. Duchesne, and J. S. Mao. 2005. Wolves influence elk movements: behavior shapes a trophic cascade in Yellowstone National Park. *Ecology* 86: 1320–1330.

Fryxell, J. M. 1987. Food limitation and demography of a migratory antelope, the white-eared kob. *Oecologia* 72: 83–91.

Fuller, J. A., R. A. Garrott, P. J. White, K. E. Aune, T. J. Roffe, and J. C. Rhyan. 2007. Reproduction and survival of Yellowstone bison. *Journal of Wildlife Management* 71: 2365–2372.

Gaillard, J.-M., J. M. Boutin, D. Delorme, G. van Laere, P. Duncan, and J. D. Lebreton. 1997. Early survival in roe deer: causes and consequences of cohort variation in two contrasted populations. *Oecologia* 112: 502–513.

Garrott, R. A., L. L. Eberhardt, P. J. White, and J. Rotella. 2003. Climate-induced variation in vital rates of an unharvested large-herbivore population. *Canadian Journal of Zoology* 81: 33–45.

Georgiadis, N., M. Hack, and K. Turpin. 2003. The influence of rainfall on zebra population dynamics: implications for management. *Journal of Applied Ecology* 40: 125–136.

Gonzalez, G. and J.-P. Crample. 2001. Mortality patterns in a protected population of isards. *Canadian Journal of Zoology* 79: 2072–2079.

Grenfell, B. T., K. Wilson, B. F. Finkenstadt, et al. 1998. Noise and determinism in synchronized sheep dynamics. *Nature* 394: 674–677.

Grotan, V., B.-E. Saether, S. Engen, et al. 2005. Climate causes large-scale spatial synchrony in population fluctuations of a temperate herbivore. *Ecology* 86: 1472–1482.

Hallett, T. B., T. Coulson, J. G. Pilkington, T. H. Clutton-Brock, J. M. Pemberton, and B. T. Grenfell. 2004. Why large-scale climate indices seem to predict ecological processes better than local weather. *Nature* 430: 71–74.

Hebblewhite, M. 2005. Predation by wolves interacts with North Pacific Oscillation (NPO) on a western North American elk population. *Journal of Animal Ecology* 74: 226–233.

Helle, T. and I. Kojola. 2008. Demographics in an alpine reindeer herd: effects of density and winter weather. *Ecography* 31: 221–230.

Hillman, J. C. and A. K. K. Hillman. 1977. Mortality of wildlife in Nairobi National Park, during the drought of 1973–74. *East African Wildlife Journal* 15: 1–18.

Hone, J. and T. H. Clutton-Brock. 2007. Climate, food, density and wildlife population growth rate. *Journal of Animal Ecology* 76: 361–367.

Jacobson, A. R., A. Provenzale, A. von Hardenerg, B. Bassano, and M. Festa-Bianchet. 2004. Climate forcing and density dependence in a mountain ungulate population. *Ecology* 85: 1598–1610.

Keep, M. E. 1973. Factors contributing to a population crash of nyala in Ndumu Game Reserve. *Lammergeyer (Natal)* 19: 16–23.

Klein, D. R. 1991. Caribou in the changing north. *Applied Animal Behavior Science* 29: 279–291.

Knight, M. H. 1995. Drought-related mortality of wildlife in southern Kalahari and the role of man. *African Journal of Ecology* 33: 377–394.

Krebs, C. J. and D. Berteaux. 2006. Problems and pitfalls in relating climatic variability to population dynamics. *Climate Research* 32: 143–149.

Langvatn, R., S. D. Albon, T. Burkey, and T. H. Clutton-Brock. 1996. Climate, plant phenology and variation in age of first reproduction in a temperate herbivore. *Journal of Animal Ecology* 65: 653–670.

Laundre, J. W., L. Hernadez, and K. B. Altendorf. 2001. Wolves, elk and bison: reestablishing the "landscape of fear" in Yellowstone National Park, U.S.A. *Canadian Journal of Zoology* 79: 1401–1409.

Lima, M. and A. Berryman. 2006. Predicting non-linear and non-additive effects of climate: the Alpine ibex revisited. *Climate Research* 32: 129–135.

Loison, A. and R. Langvatn. 1998. Short- and long-term effects of winter and spring weather on growth and survival of red deer in Norway. *Oecologia* 116: 489–500.

Lubow, B. C., F. J. Singer, T. L. Johnson, and D. C. Bowden. 2002. Dynamics of interacting elk populations within and adjacent to Rocky Mountain National Park. *Journal of Wildlife Management* 66: 757–775.

Lubow, B. C. and B. L. Smith. 2004. Population dynamics of the Jackson elk herd. *Journal of Wildlife Management* 68: 810–829.

Marchant, R., C. Mumbi, S. Behera, and T. Yamagata. 2007. The Indian Ocean dipole–the unsung driver of climatic variability in East Africa. *African Journal of Ecology* 45: 4–16.

Marshal, J. P., P. R. Krausman, V. C. Bleich, W. B. Ballard, and J. D. McKeever. 2002. Rainfall, El Nino, and the dynamics of mule deer in the Sonoran desert, California. *Journal of Wildlife Management* 66: 1283–1289.

McKinney, T., T. W. Smith, and J. C. de Vos. 2006. Evaluation of Factors Potentially Influencing A Desert Bighorn Sheep Population. *Wildlife Monographs* No. 164, 36 pp.

McPhaden, M. J. S. E. Zebiak, and M. H. Glantz. 2006. ENSO as an integrating concept in earth science. *Science* 314: 1740–1745.

Mduma, S. A. R., A. R. E. Sinclair, and R. Hilborn. 1999. Food regulates the Serengeti wildebeest: a 40-year record. *Journal of Animal Ecology* 68: 1101–1122.

Mech, L. D., R. E. McRoberts, R. O. Peterson, and R. E. Page. 1987. Relationship of deer and moose populations to previous winter's snow. *Journal of Animal Ecology* 56: 615–627.

Mills, M. G. L., H. C. Biggs, and I. J. Whyte. 1995. The relationship between rainfall, lion predation and population trends in African herbivores. *Wildlife Research* 22: 75–88.

Milner, J. M., D. A. Elston, and S. D. Albon. 1999. Estimating the contributions of population density and climatic fluctuations to interannual variation in survival of Soay sheep. *Journal of Animal Ecology* 68: 1235–1247.

Murray, D. L., E. W. Cox, W. B. Ballard, et al. 2006. Pathogens, Nutritional Deficiency, and Climatic Influences on a Declining Moose Population. *Wildlife Monographs* No. 166, 30pp.

Mysterud, A., R. Langvatn, N. G. Yoccoz, and N. C. Stenseth. 2001. Plant phenology, migration and geographic variation in body weight of a large herbivore: the effect of variable topography. *Journal of Animal Ecology* 70: 915–923.

Novellie. P. 1986. Relationship between rainfall, population density and the size of the bontebok lamb crop in the Bontebok National Park. *South African Journal of Wildlife Research* 16: 39–46.

Ogutu, J. O. and N. Owen-Smith. 2003. ENSO, rainfall and temperature influences on extreme population declines among African savanna ungulates. *Ecology Letters* 6: 412–419.

Ogutu, J. O. and N. Owen-Smith. 2005. Oscillations in large herbivore populations: are they related to predation or rainfall? *African Journal of Ecology* 43: 332–339.

Ogutu, J. O., H.-P. Piepho, H. T. Dublin, N. Bhola, and R. S. Reid. 2008a. Rainfall influences on ungulate population abundance in the Mara-Serengeti ecosystem. *Journal of Animal Ecology* 77: 814–829.

Ogutu, J. O., H.-P. Piepho, H. T. Dublin, N. Bhola, and R. S. Reid. 2008b. El Nino-Southern Oscillation, rainfall, temperature and Normalized Difference Vegetation Index fluctuations in the Mara-Serengeti ecosystem. *African Journal of Ecology* 46: 132–143.

Ottersen, G., B. Planque, A. Belgrano, E. Post, P. C. Reid, and N. C. Stenseth. 2001. Ecological effects of the North Atlantic Oscillation. *Oecologia* 128: 1–14.

Owen-Smith, N. 1990. Demography of a large herbivore, the greater kudu *Tragelaphus strepsiceros*, in relation to rainfall. *Journal of Animal Ecology* 59: 893–913.

Owen-Smith, N. 2000. Modeling the population dynamics of a subtropical ungulate in a variable environment: rain, cold and predators. *Natural Resource Modeling* 13: 57–87.

Owen-Smith, N. 2006. Demographic determination of the shape of density dependence for three African ungulate populations. *Ecological Monographs* 76: 93–109.

Owen-Smith, N. D. R. Mason, and J. O. Ogutu. 2005. Correlates of survival rates for 10 African ungulate populations: density, rainfall and predation. *Journal of Animal Ecology* 74: 774–788.

Owen-Smith, N., and M. G. L. Mills. 2006. Manifold interactive influences on the population dynamics of a multispecies ungulate assemblage. *Ecological Monographs* 76: 73–92.

Owen-Smith, N. and M. G. L. Mills. 2008. Shifting prey selection generates contrasting herbivore dynamics within a large-mammal predator-prey web. *Ecology* 89: 1120–1133.

Patterson, B. R. and V. A. Power. 2002. Contributions of forage competition, harvest, and climate fluctuations to changes in population growth of northern white-tailed deer. *Oecologia* 130: 62–71.

Peek, J. M., B. Dennis, and T. Herschey. 2002. Predicting population trends of mule deer. *Journal of Wildlife Management* 66: 729–736.

Pettorelli, N., A. Mysterud, N. G. Yoccoz, R. Langvatn, and N. C. Stenseth. 2005b. Importance of climatological downscaling and plant phenology for red deer in heterogeneous landscapes. *Proceedings of the Royal Society of London B: Biological Sciences* 272: 2357–2364.

Pettorelli, N., F. Pelletier, A. von Hardenberg, M. Festa-Bianchet, and S. D. Cote. 2007. Early onset of vegetation growth vs. rapid greenup: impacts on juvenile ungulates. *Ecology* 88: 381–390.

Pettorelli, N., R. B. Weladji, O. Holand, A. Mysterud, H. Breie, and N. C. Stenseth. 2005a. The relative role of winter and spring conditions linking climate and landscape-scale plant phenology to alpine reindeer performance. *Biology Letters* 1: 24–26.

Portier, C., M. Festa-Bianchet, J.-M. Gaillard, J. T. Jorgenson, and N. G. Yoccoz. 1998. Effect of density and weather on survival of bighorn lambs. *Journal of Zoology* 245: 271–278.

Post, E. and M. C. Forchhammer. 2002. Synchronization of animal population dynamics by large-scale climate. *Nature* 420: 168–171.

Post, E. and N. C. Stenseth. 1998. Large-scale climatic fluctuation and population dynamics of moose and white-tailed deer. *Journal of Animal Ecology* 67: 537–543.

Post, E. and N. C. Stenseth. 1999. Climatic variability, plant phenology, and northern ungulates. *Ecology* 80: 1322–1339.

Reimers, E. 1995. Rangifer population ecology: a Scandinavian perspective. *Rangifer* 17: 105–118.

Royama, T. 1992. *Analytical Population Dynamics.* Chapman & Hall, London.

Sand, H. 1996. Life history patterns in female moose: the relationship between age, body size, fecundity and environmental conditions. *Oecologia* 106: 212–220.

Sauer, J. R. and M. S. Boyce 1983. Density dependence and survival of elk in southwestern Wyoming. *Journal of Wildlife Management* 47: 31–37.

Simpson, D. C., L. A. Harveson, C. E. Brewer, R. E. Walser, and A. R. Sides. 2007. Influence of precipitation on pronghorn demography in Texas. *Journal of Wildlife Management* 71: 906–910.

Sinclair, A. R. E. and Arcese, P. 1995. Population consequences of predation-sensitive foraging: the Serengeti wildebeest. *Ecology* 76: 882–891.

Sinclair, A. R. E., S. A. R. Mduma, J. G. C. Hopcraft, J. M. Fryxell, R. Hilborn, and S. Thirgood. 2007. Long-term ecosystem dynamics in the Serengeti: lessons for conservation. *Conservation Biology* 21: 580–590.

Skogland, T. 1985. The effects of density dependent resource limitations on the demography of wild reindeer. *Journal of Animal Ecology* 54: 359–374.

Smuts, G. L. 1978. Interrelations between predators, prey, and their environment. *BioScience* 28: 316–320.

Solberg, E. J., P. Jordhoy, O. Strand, et al. 2001. Effects of density-dependence and climate on the dynamics of a Svalbard reindeer population. *Ecography* 24: 441–451.

Solberg, E. J., B.-E. Saether, O. Strand, and A. Loison. 1999. Dynamics of a harvested moose population in a variable environment. *Journal of Animal Ecology* 68: 186–204.

Spinage, C. A. and J. M. Matlhare. 1992. Is the Kalahari cornucopia fact or fiction? A predictive model. *Journal of Applied Ecology* 29: 605–610.

Stenseth, N. C., K.-S. Chan, G. Tavecchia, et al. 2004. Modelling non-additive and non-linear signals from climatic noise in ecological time series: Soay sheep as an example. *Proceedings of the Royal Society of London B: Biological Sciences* 271: 1985–1993.

Stenseth, N. C. and A. Mysterud. 2005. Weather packages: finding the right scale and composition of climate in ecology. *Journal of Animal Ecology* 74: 1195–1198.

Stenseth, N. C., G. Ottersen, J. W. Hurrell, et al. 2003. Studying climate effects on ecology through the use of climate indices: the North Atlantic Oscillation, El Nino Southern Oscillation and beyond. *Proceedings of the Royal Society of London B: Biological Sciences* 270: 2087–2096.

Turner, M. G., Y. Wu, L. L. Wallace, W. H. Romme, and A. Bronkert. 1994. Simulating winter interactions among ungulates, vegetation, and fire in northern Yellowstone Park. *Ecological Applications* 4: 472–496.

Tyler, N. J. C., M. C. Forchhammer, and N. A. Oritsland. 2008. Nonlinear effects of climate and density on the dynamics of a fluctuation population of reindeer. *Ecology* 89: 1675–1686.

Tyson, P. D. 1991. Climate change in southern Africa: past and present conditions and possible future scenarios. *Climate Change* 18: 241–258.

Tyson, P. D. and C. K. Gatebe. 2001. The atmosphere, aerosols, trace gases and biogeochemical change in southern Africa: a regional integration. *South African Journal of Science* 97: 106–118.

Vucetich, J. A., D. W. Smith, and D. R. Stahler. 2005. Influence of harvest, climate and wolf predation on Yellowstone elk, 1961–2004. *Oikos* 111: 259–270.

Walker, B. H., R. H. Emslie, N. Owen-Smith, and R. J. Scholes. 1987. To cull or not to cull: lessons from a southern African drought. *Journal of Applied Ecology* 24: 381–402.

Wang, G. M., N. T. Hobbs, F. J. Singer, D. S. Ojima, and B. C. Lubow. 2002. Impacts of climate change on elk population dynamics in Rocky Mountain National Park, Colorado, USA. *Climate Change* 54: 205–223.

Wang, G., N. T. Hobbs, S. Twombly, et al. 2009. Density dependence in northern ungulates: interactions with predation and resources. *Population Ecology* 51: 123–132.

Weladji, R. B. and O. Holand. 2006. Influences of large-scale climatic variability on reindeer population dynamics: implications for reindeer husbandry in Norway. *Climate Research* 32: 119–127.

Weladji, R. B., D. R. Klein, O. Holand, and A. Mysterud. 2002. Comparative response of *Rangifer tarandus* and other northern ungulates to climatic variability. *Rangifer* 22: 33–50.

White, P. J. and R. A. Garrott. 2005. Yellowstone's ungulates after wolves–expectations, realizations and predictions. *Biological Conservation* 125: 141–152.

Wilmers, C. C., E. Post, and A. Hastings. 2007. The anatomy of predator-prey dynamics in a changing climate. *Journal of Animal Ecology* 76: 1037–1044.

Wright, G. J., R. O. Peterson, D. W. Smith, and T. O. Lemke. 2006. Selection of northern Yellowstone elk by grey wolves and hunters. *Journal of Wildlife Management* 70: 1070–1078.

4

Demographic processes: lessons from long-term, individual-based studies

Jean-Michel Gaillard[1], Tim Coulson[2] and
Marco Festa-Bianchet[3]

[1]*Laboratoire de Biometrie et Biologie Evolutive,
Universite Claude Bernard Lyon, Villeurbanne Cedex, France*

[2]*Department of Life Sciences, Imperial College London,
Ascot, United Kingdom*

[3]*Department of Biology, University of Sherbrooke,
Sherbrooke, Canada*

Two approaches have been widely used to study the population dynamics of large herbivores. One, referred to as either the pattern-oriented approach (Coulson et al. 2000) or the density paradigm (Krebs 2002), consists of statistically summarizing time series of population counts and inferring how density and environmental factors such as climate, contribute to changes in population size or growth. The second approach, called either the process-oriented approach (Coulson et al. 2000) or the mechanistic paradigm (Krebs 2002), involves the estimation of age-specific demographic parameters such as survival and reproduction from long-term monitoring of recognizable individuals and the calculation of how each demographic parameter affects population growth. Advances in the analysis of time series, and the increasing availability of yearly counts of populations of large herbivores, have led to an explosion in the use of pattern-oriented studies in the last decade. These studies suggest that most populations of large

Dynamics of Large Herbivore Populations in Changing Environments, 1st edition. Edited by Norman Owen-Smith.
© 2010 Blackwell Publishing

herbivores show direct or slightly delayed density dependence, and respond negatively to harsh weather ranging from snow to drought (Chapter 3). However, population counts seldom allow distinctions among age classes apart from the young of the year (Bonenfant et al. 2005), so that changes in age structure, an important feature of long-lived and iteroparous species including large herbivores, cannot be assessed. As a result, the explanatory power of time series analyses is often limited, as demonstrated by the Soay sheep case study (Coulson et al. 2001; see Chapter 1).

Recent research on population dynamics of large herbivores suggests that an accurate assessment of demographic processes is best obtained with data from the monitoring of known-aged individuals, which unfortunately are difficult to obtain. Based on long-term studies of known-age marked individuals, mostly restricted to the holarctic zone, we will (i) present the main life-history features of large herbivores, (ii) evaluate how different demographic parameters affect their population growth rate, and (iii) explore complex interactions between climatic variation, density-dependence, and individual variation.

4.1   Life history of large herbivores: a brief review

Despite a high diversity of lifestyles, large herbivores are a rather homogeneous group from a life-history viewpoint. Empirical studies in the last 20 years have repeatedly reported that life-history strategies vary along a fast–slow gradient, from short-lived species that reproduce early and produce many offspring at each of their few breeding attempts to long-lived species that reproduce late and produce a few offspring at each of their numerous breeding attempts (Stearns 1983, Gaillard et al. 1989, Bielby et al. 2007). Large herbivores are clearly found closer to the slow end of the continuum. The ranking of species persists after correcting for allometric constraints that affect most life-history traits (Stearns 1983).

Compared to most other mammals of similar size, a typical large herbivore female is highly precocious as a neonate, with a high birth mass relative to its mother's mass (26% higher than in other mammals, Robbins and Robbins 1979) and rapid early growth (Robbins and Robbins 1979). It is closely associated with its mother during its first year of life but can survive independently after weaning (which typically occurs within one year after birth). The young female settles in its own home range between 1 and 3 years of age but often remains close to its mother for its entire life. It grows to more than 80% of its asymptotic size before its first mating, gives birth for the first time at 2 or 3 years of age, and produces generally one offspring per breeding attempt, usually each year. After reaching maturity it experiences high survival and reproductive output during a prime-age stage between 2 or 3 and 8 or 10 years of age, and thereafter enters a senescent stage during which survival and reproductive performance gradually decrease with increasing age.

Hence the age of the individual has an important effect on both survival probability and reproductive output. Survival from birth to 1 year is generally low and highly variable over time (overall mean of 0.52 (N = 51 populations), CV of 0.35 (N = 43), Gaillard et al. 2000) and reproduction in the first year is exceedingly rare. In contrast, prime-aged females show a mean annual survival rate of 0.87 (N = 57 populations), typically with little variation between years (CV of 0.09, N = 48; Gaillard et al. 2000), and high fecundity (overall mean proportion of females that give birth in a given year 0.82, N = 59, with CV = 0.13, N = 51; Gaillard et al. 2000).

Age-specific variation in demographic parameters means that individuals spend longer in some demographic classes than in others. For example, once an individual has grown from a neonate to a sexually mature adult it has passed through a risky period when expected survival is low. It then enters a period when it is in rather good condition under most environmental conditions but the poorest, and survival is generally high. Because this period lasts several years, a large proportion of the population is in this class. Once individuals begin to senesce, their survival decreases, and they are not expected to remain in the senescent class for as long as they remain in the prime-aged adult class. As a consequence, age structure typically describes variation in demographic parameters of large herbivores better than classifications based on mass. However, within an age-class, body mass can independently influence demographic rates. For example, large and heavy adult females most often reproduce earlier, have larger litter sizes, produce higher quality offspring, and live longer than light ones (Gaillard et al. 2000). In other words, while individual differences in body mass matter, the age-specific differences in survival and fertility observed in large ungulates are not simply due to age-related differences in mass.

A life-cycle graph appropriate for demographic analyses of populations of large herbivores is presented in Fig. 4.1. Based upon this life-cycle graph, a demographic analysis involves two steps. First, age-specific demographic parameters are estimated from monitoring of known-aged individuals. For bovids such as ibex, bighorn sheep or chamois, annual horn increments provide a reliable way of aging animals (e.g., Schröder and Elsner-Schack 1985) but for most species, age estimates of live adult animals are imprecise (e.g., Hewison et al. 1999). Therefore, the most reliable way of monitoring known-aged individuals is to mark them as juveniles. Longitudinal monitoring is ideally required to account for individual variation in quality, which arises from inter-cohort variation in environmental conditions

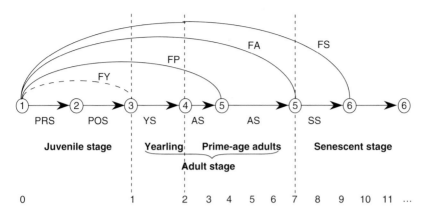

Figure 4.1 A life cycle graph for demographic analyses of the female segment of populations of large herbivores. (1) Newborn, (2) Weaned young, (3) Yearlings, (4) Two-year-olds, (5) Prime-aged adults, (6) Senescent adults (older than 7 years), PRS: summer survival of juveniles; POS: winter survival of juveniles; YS: yearling survival; AS: prime-age adult survival; SS: senescent survival; FY: fecundity of yearlings; FP: fecundity of 2-years-olds; FA: fecundity of prime-aged adults; FS: fecundity of senescent adults. The figure shows survival from one age group to the next (*straight lines*) and reproduction of a given age group (*curved lines*).

(e.g., climate and population density) around the time of birth, spatial differences in habitat quality, and variation in genotypic and phenotypic attributes. Individual variation may decrease with increasing age because of the filtering effects of nonrandom mortality (Vaupel et al. 1979). Because nonrandom mortality leads to an increase in average individual quality with age, it may bias estimates of age-specific parameters. As a result, transversal data may increasingly overestimate individual performance with increasing age and underestimate the magnitude of senescence, or even fail to detect senescence completely (Cam et al. 2002).

Even with longitudinal monitoring of known-aged individuals, it is rarely possible to obtain complete lifetime data for all marked individuals, because in many studies not all surviving animals are detected every year. Therefore, statistical procedures are required to account for variability in detection probability and avoid bias in survival estimates (Nichols 1992). Capture-mark-recapture methods (Lebreton et al. 1992) have been used to estimate age-specific survival in large herbivores, including roe deer (Gaillard et al. 1993), chamois (Loison et al. 1994), bighorn sheep (Jorgenson et al. 1997), mouflon (Cransac et al. 1997), moose (Stubsjoen et al. 2000), Soay sheep (Catchpole et al. 2000), white-tailed deer (DelGiudice et al. 2002), elk (Garrott et al. 2003), red deer (Catchpole et al. 2004), and ibex (Toïgo et al. 2007). Capture-mark-recapture models are also useful

to estimate reproductive parameters such as age-specific breeding probabilities (Clobert et al. 1994). The recent development of multistate models
(Lebreton and Pradel 2002) has provided a general way to estimate directly
the transition probabilities among all categories of stage-structured models
(Nichols et al. 1992, Fujiwara and Caswell 2002) and could also estimate
covariance among demographic rates.

Once estimates of all demographic parameters have been obtained (generally at a yearly interval, because most large herbivores only have one
reproductive attempt per year), a second step involves a demographic analysis, to assess how the mean and variance of each parameter influence population growth, variation in population growth, or the population dynamics.
This analysis requires a measure of population performance. While studies
dealing with conservation issues are generally based on population viability analysis (e.g., Boyce 1992), focusing on the probability of extinction
(e.g., Hosack et al. 2002), most demographic analyses have addressed population growth, especially after Caswell's (2001) comprehensive review of
the different ways to quantify population growth and its variation. The
measure that is most widely used in demographic analyses is the asymptotic deterministic growth rate (denoted λ) obtained as the greatest positive
eigenvalue of the Leslie matrix including age-specific fertilities (on the first
row) and age-specific survival (on the first sub-diagonal). The suitability
of λ as a metric for demographic analyses, however, has been challenged (Benton and Grant 2000, Grant and Benton 2000, 2003) because
under density-dependence, populations are not expected to increase in size
indefinitely. Although population size could be used instead of λ in such
situations, as pointed out by Caswell (2001), even populations with a λ
of 1 should show fluctuations in population size with time in response to
changes in climatic conditions, habitat quality, the prevalence of diseases,
or predation pressure. Therefore, the yearly growth fluctuates around 1, so
that the relative contributions of changes in demographic parameters on
the variance in observed population growth can be estimated.

Most demographic analyses of large herbivores have been deterministic, as may be justified by the relatively low temporal variation
in demographic parameters (Benton and Grant 1996). Very recent
stochastic analyses (Ezard et al. 2008, Morris et al. 2008) suggest that
deterministic measures provide an acceptable ranking of the contributions
of demographic parameters to the long-run population growth rate, but
could miss important demographic processes. In density-independent
environments, the asymptotic deterministic growth rate usually
overestimates the long-run growth rate (e.g., Tuljapurkar 1989). The
long-run population growth rate, also called the stochastic growth
rate (noted a, see e.g. Tuljapurkar 1989 or λ_s, see e.g. Caswell 2001)

provides a more appropriate measure of the actual population growth rate. However, analyses to date suggest that similar conclusions are obtained from decompositions of both measures. When density dependence occurs, λ and λ_s both effectively have a value of 1. In the following section, we use the deterministic elasticities and sensitivities of demographic parameters, which provide reliable approximation of actual measures in populations of large herbivores.

4.2 Differential contributions of demographic parameters to population growth

Eberhardt (1991) was among the first to highlight the heuristic value of matrix models in the context of population dynamics of large herbivores. Since then, most demographic analyses of populations of large herbivores sought to identify the demographic parameter that, when changed by a given absolute or relative amount, will lead to the largest change in the population growth rate (Gaillard and Yoccoz 2003). Such prospective analyses (*sensu* Caswell 2000) can rank demographic parameters according to the impact that a given change in each parameter may have on mean population growth rate if that parameter varied independently of variation in any other demographic rates. The impact can be measured on different scales. Sensitivity, which is the partial derivative of λ to a change of a given demographic parameter, measures the absolute impact, i.e. the change in population growth caused by changing each parameter independently by a given amount; while elasticity, which is the partial derivative of λ to a change of a given demographic parameter on a log-scale, measures the relative impact, i.e. the change in population growth caused by changing each parameter independently by a given proportion. As most parameters in populations of large herbivores are scaled between 0 and 1 (except for litter size, a component of fecundity that can reach 3 in a few cases and up to 10 in some suid species, Hayssen et al. 1993), choosing sensitivity or elasticity has little effect on the ranking.

All prospective studies so far have reported that adult female survival is the key demographic parameter in populations of large herbivores, regardless of whether sensitivities or elasticities are used (see Gaillard et al. 2000 for a review). The conclusion that adult female survival has the highest elasticity of any demographic parameter is likely to be correct for any population of large herbivore. Indeed, elasticities of population growth to changes in all age-specific survival sum to 1, so that changing all survival rates from birth to the oldest age by the same proportional amount will change the population growth by the same amount. In other words, the elasticity of population growth to survival after 1 year of age can be measured as one minus the elasticity of population growth to juvenile survival. Moreover, if juvenile survival is independent of maternal age, the elasticity of population growth to juvenile survival corresponds to the summed elasticities of all age-specific fecundity terms (i.e., the first row elements of the Leslie matrix). Under these conditions, generation time can be estimated as the inverse of the elasticity of population growth to juvenile survival (Brooks and Lebreton 2001). It follows that a population should have a generation time shorter than 2 years for population growth rate to have a lower elasticity to adult survival than to juvenile survival or to fecundity. Such a short generation time would require a mean age of mothers of less than 2 years, which is clearly impossible for unhunted populations of large herbivores. Indeed, based on the allometric relationship between generation time and body mass in large herbivores, Gaillard et al. (2008) reported that no extant species of large herbivore is small enough to have such a short generation time.

The ability of elasticity analyses to identify key demographic parameters in real populations has been challenged (Mills et al. 1999), because demographic parameters do not vary over time by the same amount. In large herbivore populations, recruitment parameters such as juvenile survival, age-specific breeding proportions, and mean litter size (for polytocous species) vary much more than adult female survival (Gaillard et al. 1998, 2000, Eberhardt 2002). Thus, instead of measuring the response of population growth to the same given change of demographic parameters (as performed in elasticity analyses), one could look for the demographic parameter with observed variation over time contributing the most to the observed variation in population growth.

Key-factor analysis was designed with that aim and was applied to some populations of large herbivores by earlier studies (Sinclair 1977, Clutton-Brock et al. 1985). This approach has been criticized (Manly 1979, Royama 1996) and is now generally replaced by perturbation analyses based on Leslie matrices (Caswell 2000). These methods involve perturbing a model parameter and examining consequences on model predictions. When complete yearly monitoring of all individuals in the population can be achieved, an exact decomposition of variance in observed population growth can be obtained using a method called *structured demographic accounting* (SDA) (Brown and Alexander 1991). This technique has been applied to red deer (Brown et al. 1993, Albon et al. 2000, Coulson et al. 2005), Soay sheep (Coulson et al. 1999), and bighorn sheep (Coulson et al. 2005). However, in most cases only yearly estimates of demographic parameters are available so that an SDA cannot be performed.

Retrospective analyses (*sensu* Caswell 2000), like key-factor analysis or SDA, provide an approximate decomposition of the observed temporal variance in population growth into contributions from the observed temporal variation in demographic rates. Early population analyses of large herbivores (Gaillard et al. 2000, 2003a) only included the main contributions of demographic parameters and reported that those with the highest potential impact on population growth rate as measured by elasticity or sensitivity are not necessarily those responsible for observed variation in growth rate over a time window of about 10 years. Despite its low potential impact on population growth for a given proportional change, variation in juvenile survival had a larger contribution to observed temporal variation in population growth rate than variation in prime-age adult survival in roe deer, bighorn sheep, and red deer (Gaillard et al. 2000). Thus, when variation in population growth of large herbivores is analyzed, it is generally found that the largest additive contribution comes from variation in juvenile survival.

More recently, however, comprehensive SDA analyses of red deer, Soay sheep, and bighorn sheep populations have shown that covariation among demographic parameters generated by similar responses to environmental signals such as extreme climatic events, epizootics, or high predation, may play an important role and should be considered. In the only matrix analysis (i.e., simply based on annual estimates of demographic parameters for a series of years) published so far on the contribution of demographic parameters (including their two-way interactions) to changes in population growth in large herbivores, Coulson et al. (2005) confirmed that there is not necessarily a link between elasticity of growth rate to a demographic parameter and the contribution of that parameter to observed variance in population growth rate. Interspecific studies of birds (Saether and Bakke 2000) and mammals (Gaillard and Yoccoz 2003) found no relationship between the elasticity of a given demographic parameter and its contribution to observed variance in population growth. Coulson et al. (2005) also showed that, when covariation among parameters is accounted for, juvenile survival does not consistently make the largest contribution to changes in population growth. Therefore, when nonadditive effects are incorporated into the demographic analysis, the picture can change. The demographic parameter that contributes the most to observed variation in population growth of large herbivores is context-dependent and thereby difficult to predict. The latest decompositions have focused on identifying contributions of density-dependence, climate, and age-structure fluctuations to population dynamics (Lande et al. 2006, Coulson et al. 2008).

4.3   Climatic variation, density-dependence, and individual variability

As discussed above, estimating age-specific demographic parameters and their variability over time provides the essential information required to perform demographic analyses. However, while such analyses may assess the relative importance of demographic parameters on different scales (absolute vs. proportional change), or using alternative measures (mortality vs. survival), and can identify conservation or management responses needed, they tell us nothing about what ecological processes generate variation. The same value of survival or reproductive trait in two different years can be caused by different processes. We usually recognize three major sources of variation in demographic parameters (Lande et al. 2003): environmental variation, density-dependence, and demographic stochasticity. Recent advances in population dynamics of vertebrates allow us to add a fourth component: frequency-dependent effects generated by fluctuating age-structure (e.g., Coulson et al. 2006, Lande et al. 2006).

Environmental variation in climate, habitat quality, disease, abundance of predators, parasites and competitors, and in human influences can have strong effects on populations of large herbivores. The effects of environmental variation on demographic parameters have been well documented in populations of large herbivores, especially for juvenile survival (Chapter 3). Among the factors causing variation in demographic parameters are changes in winter harshness as measured by seasonal weather metrics such as the winter North Atlantic Oscillation (NAO) (Coulson et al. 2001, Weladji et al. 2002, Catchpole et al. 2004) or the Arctic Oscillation (AO) (Aanes et al. 2002), or by snow depth (Adams et al. 1995, DelGiudice et al. 2006); changes in spring–summer temperature or precipitation (Coughenour and Singer 1996, Gaillard et al. 1997, Portier et al. 1998, Garel et al. 2004); spatial heterogeneity in habitat quality (Saether and Heim 1993, Coulson et al. 1997, Coulson et al. 1999, Pettorelli et al. 2003, 2005, McLoughlin et al. 2006); outbreaks of disease (Loison et al. 1996, Cransac et al. 1997, Jorgenson et al. 1997); changes in predation (Linnell et al. 1995, Byers 1997, Kjellander et al. 2004, Wittmer et al. 2005, 2007, Festa-Bianchet et al. 2006); and changes in hunting pressure (Langvatn and Loison 1999).

Density-dependence is pervasive in large herbivores. Reviews of density dependence in large mammals (Fowler 1987, Bonenfant et al. 2009) mostly support Eberhardt's (1977) hypothesis that demographic parameters respond to changes in density in a sequential order, with juvenile survival reacting first, then age at first reproduction, then reproductive output of prime-aged females, and lastly prime-age survival. In very large species, the age at first reproduction might react to changes in density before juvenile survival, as reported for elephants (Wittemyer et al. 2007). Density-dependent responses interact with environmental variation, so that demographic parameters change much more under harsh than under good conditions for a given change in density (Portier et al. 1998, Coulson et al. 2001, Solberg et al. 2001, Boyce et al. 2006, but see Hone and Clutton-Brock 2007).

Unlike the impacts of environmental factors and density-dependence, the effects of demographic stochasticity in populations of large herbivores have rarely been quantified, although they can be important in small populations, as reported for moose (Saether et al. 2007). However, in large populations, the influence of demographic stochasticity is expected to be much smaller than that of environmental variation.

Assessing the influences of environmental conditions, density-dependence, and demographic stochasticity on population dynamics of large herbivores by comparing yearly estimates of demographic parameters with a series of covariates, as done in most studies, does not provide a complete understanding of demographic processes. A typical population of large herbivores will include several cohorts (often more than 10), and adult females may overlap temporally with mothers and grandmothers. A given environmental event (either good or bad) will affect demographic rates differentially according to the age of the individuals. By affecting the youngest age class most heavily, that event will also influence in a predictable way the size and age–sex structure of the population in later years, often with long-lasting effects on individuals. The best evidence for complex and long-term effects of changes occurring in a given year involves cohort effects, which have been well studied in populations of large herbivores (Albon et al. 1987, Gaillard et al. 1997, Forchhammer et al. 2001, Solberg et al. 2004). A high proportion of the juveniles born in a good year will be recruited into the population as adults, whereas in a poor year few juveniles will survive, generating numerical cohort effects. When populations are resource-limited by high density or poor habitat quality, individuals born in adverse years that survive to adulthood may show reduced demographic performance throughout life. Such quality effects of cohort variation (*sensu* Gaillard et al. 2003b) have been reported in several life history traits including adult body mass (Pettorelli et al. 2002), senescence rates in both reproduction and survival (Nussey et al. 2007), and lifetime reproductive success (Rose et al. 1998).

Alternatively, long-lasting effects of being born in a poor year could be counterbalanced by strong viability selection during the first year. If the very high early mortality of juveniles during the poorest years affected mainly individuals with lower phenotypic or genetic quality, it may produce a strong filter that should increase the average quality of adults from those cohorts, because only the best individuals would survive. Coltman et al. (1999) suggested that the most homozygous Soay sheep are more likely to die during crash years. However, while some nearly complete cohort failures have been reported (Page 1988), we know of no cases where cohorts born in a poor year outperformed good cohorts later on.

Because most large herbivores do not give birth until they are at least 2 years of age, the numerical and quality effects of cohort variation can generate high age-related variability among individuals in a given year in a population. Individuals from poor cohorts that survive to adulthood will often be lighter than those born in good years and thereby less likely to reproduce successfully at a given age. Cohort variation can thus generate markedly different average quality of prime-aged individuals. If consecutive cohorts experience a similar environment (for example, with several consecutive years of high density or unfavorable weather), consistent differences in quality may affect a large proportion of the population. Identical environmental conditions and density in two different years with different average quality of prime-aged females could then lead to different reproductive output at the population level. Therefore, the exact influence of yearly characteristics on demographic parameters will depend upon the age structure and cohort effects on individual quality. To fully understand demographic processes in large herbivores, one needs to consider inter-cohort differences in both average quality and numerical consistency, because the same environmental context will generate different population-level signatures depending on age-structure. It is therefore important to understand what cohort effects can generate fluctuations in age-structure. Identifying individual-level phenomena acting from the year of birth to the year of death (i.e., studying individual trajectories) is the only way to understand population responses to a given set of environmental conditions.

4.4   Conclusions: how can future analyses of large herbivore demography deal with complex demographic processes?

Long-term monitoring of individuals from birth to death revealed unexpected complexities in population dynamics of large herbivores. The impact of density-dependence and environmental variation varies according to the distribution of individual differences in quality. How can these complex demographic processes be accounted for by studies of population dynamics of large herbivores? We will focus on one question often asked in population biology (see Stearns 1992): what shape do the density-dependent functions affecting demographic parameters take?

The seminal paper by Gilpin and Ayala (1973) provided a starting context for empirical analyses of the shape of density-dependence. Fowler (1981, 1988) argued that long-lived species with long generation times should show the strongest density-dependence at high density, rather than conform to the logistic model with a constant strength of density-dependence over the range of densities. In contrast, after analyzing many time series of population counts, Sibly et al. (2005) proposed that density-dependence in mammals is more pronounced at low than at high density, although this finding has been criticized (e.g., Peacock and Garshelis 2006). Available empirical evidence from long-term studies of populations of large herbivores supports Fowler's viewpoint, and the sequential model (Eberhardt 1977, Gaillard et al. 2000), with an increasing number of demographic parameters showing density-dependent responses as density increases (Bonenfant et al. 2009). Eberhardt's model, however, does not tell us anything about the shape of density-dependence at the population level. We suggest that there is little to be gained by trying to predict this shape for large herbivores, because of widespread interactions between density-dependence and environmental factors in all wild populations. Population dynamics of large herbivores are never driven only by density-dependence, but are affected by a complex set of interactions among density, age structure, spatial structure, and environmental variation. Density can influence population dynamics, but does not drive population change. Therefore, the shape of density-dependence should be context-dependent, varying especially in relation to habitat quality, population structure, climatic conditions, and predation pressure.

Long-term monitoring of known-age and individually recognizable individuals has provided relevant information on demographic processes in populations of large herbivores. These studies have (i) identified which life history stage best accounts for observed variation in population growth, (ii) assessed the responses of demographic parameters to changes in population density, climate, habitat quality, predation pressure, and prevalence of diseases, and (iii) determined how the shape of density-dependent responses in demographic parameters varies according to context-specific processes. In the future, more long-term studies in tropical areas and in populations coexisting with large predators may provide insights into the effects of predation or of the predictability of habitat quality among or within years on ungulate population dynamics. Recent progress in genetic tools for studying free-ranging populations (e.g., Pemberton 2008) should allow a better understanding of how genetic and environmental factors influence variation in life history traits of large herbivores, and consequently their population dynamics and evolution. Recent studies on Soay sheep suggest that adaptive changes in free-ranging populations may be limited by contrasting responses of heritability and selection of traits to environmental variation (Wilson et al. 2006). The generality of these findings remains to be assessed.

We must, however, recognize the limitations of long-term studies of marked individuals in furthering our understanding of population biology of large herbivores. Currently, logistic problems and the intensity of fieldwork required to obtain detailed information on each individual each year make it very difficult to monitor more than about 300 marked animals at a time. That limits the spatial extent of these studies, so that questions about the importance of predation, metapopulation dynamics, and large-scale migration remain difficult to address, yet play an important role in many large herbivore populations.

Acknowledgments

We are grateful to Bernt-Erik Saether, John Fryxell, and Norman Owen-Smith for useful comments on previous drafts of this chapter.

References

Aanes, R., B. E. Saether, F. M. Smith, et al. 2002. The Arctic Oscillation predicts effects of climate change in two trophic levels in a high-arctic ecosystem. *Ecology Letters* 5: 445–453.

Adams, L. G., F. J. Singer, and B. W. Dale. 1995. Caribou calf mortality in Denali National Park, Alaska. *Journal of Wildlife Management* 59: 584–594.

Albon, S. D., T. H. Clutton-Brock, and F. E. Guinness. 1987. Early development and population dynamics in red deer. 2. Density-independent effects and cohort variation. *Journal of Animal Ecology* 56: 69–81.

Albon, S. D., T. N. Coulson, D. Brown, et al. 2000. Temporal changes in key factors and key age groups influencing the population dynamics of female red deer. *Journal of Animal Ecology* 69: 1099–1110.

Benton, T. G. and A. Grant. 1996. How to keep fit in the real world: elasticity analyses and selection pressures on life histories in a variable environment. *American Naturalist* 147: 115–139.

Benton, T. G. and A. Grant. 2000. Evolutionary fitness in ecology: comparing measures of fitness in stochastic, density-dependent environments. *Evolutionary Ecology Research* 2: 769–789.

Bielby, J., G. M. Mace, O. R. P. Bininda-Emonds, et al. 2007. The fast-slow continuum in mammalian life history: An empirical reevaluation. *American Naturalist* 169: 748–757.

Bonenfant, C., J.-M. Gaillard, T. Coulson, et al. 2009. Empirical evidence of density-dependence in populations of large herbivores. *Advances in Ecological Research* 41: 313–357.

Bonenfant, C., J.-M. Gaillard, F. Klein, et al. 2005. Can we use the young: female ratio to infer ungulate population dynamics? An empirical test using red deer Cervus elaphus as a model. *Journal of Applied Ecology* 42: 361–370.

Boyce, M. S. 1992. Population viability analysis. *Annual Review of Ecology and Systematics* 23: 481–506.

Boyce, M. S., C. V. Haridas, C. T. Lee, et al. 2006. Demography in an increasingly variable world. *Trends in Ecology and Evolution* 21: 141–148.

Brooks, E. N. and J. D. Lebreton. 2001. Optimizing removals to control a metapopulation: application to the yellow legged herring gull (*Larus cachinnans*). *Ecological Modelling* 136: 269–284.

Brown, D. and N. Alexander. 1991. The analysis of the variance and covariance of products. *Biometrics* 47: 429–444.

Brown, D., N. D. E. Alexander, R. W. Marrs, et al. 1993. Structured accounting of the variance of demographic change. *Journal of Animal Ecology* 62: 490–502.

Byers, J. A. 1997. *American Pronghorn*. University of Chicago Press, Chicago.

Cam, E., W. A. Link, E. G. Cooch, et al. 2002. Individual covariation in life-history traits: Seeing the trees despite the forest. *American Naturalist* 159: 96–105.

Caswell, H. 2000. Prospective and retrospective perturbation analyses: their roles in conservation biology. *Ecology* 81: 619–627.

Caswell, H. 2001. *Matrix Population Models: Construction, Analysis, and Interpretation*. Sinauer, Sunderland.

Catchpole, E. A., Y. Fan, B. J. T. Morgan, et al. 2004. Sexual dimorphism, survival and dispersal in red deer. *Journal of Agricultural Biological and Environmental Statistics* 9: 1–26.

Catchpole, E. A., B. J. T. Morgan, T. N. Coulson, et al. 2000. Factors influencing Soay sheep survival. *Journal of the Royal Statistical Society Series C: Applied Statistics* 49: 453–472.

Clobert, J., J. D. Lebreton, D. Allaine, et al. 1994. The estimation of age-specific breeding probabilities from recaptures or resightings in vertebrate populations. 2. Longitudinal models. *Biometrics* 50: 375–387.

Clutton-Brock, T. H., M. Major, and F. E. Guinness. 1985. Population regulation in male and female red deer. *Journal of Animal Ecology* 54: 831–846.

Coltman, D. W., J. G. Pilkington, J. A. Smith, and J. M. Pemberton. 1999. Parasite-mediated selection against inbred Soay sheep in a free-living, island population. *Evolution* 53: 1259–1267.

Coughenour, M. B. and F. J. Singer. 1996. Elk population processes in Yellowstone National Park under the policy of natural regulation. *Ecological Applications* 6: 573–593.

Coulson, T., S. Albon, F. Guinness, et al. 1997. Population substructure, local density, and calf winter survival in red deer (*Cervus elaphus*). *Ecology* 78: 852–863.

Coulson, T., S. Albon, J. Pilkington, et al. 1999. Small-scale spatial dynamics in a fluctuating ungulate population. *Journal of Animal Ecology* 68: 658–671.

Coulson, T., T. G. Benton, P. Lundberg, et al. 2006. Estimating individual contributions to population growth: evolutionary fitness in ecological time. *Proceedings of the Royal Society B: Biological Sciences* 273: 547–555.

Coulson, T., E. A. Catchpole, S. D. Albon, et al. 2001. Age, sex, density, winter weather, and population crashes in Soay sheep. *Science* 292: 1528–1531.

Coulson, T., T. H. G. Ezard, F. Pelletier, et al. 2008. Estimating the functional form for the density dependence from life history data. *Ecology* 89: 1661–1674.

Coulson, T., J.-M. Gaillard, and M. Festa-Bianchet. 2005. Decomposing the variation in population growth into contributions from multiple demographic rates. *Journal of Animal Ecology* 74: 789–801.

Coulson, T., E. J. Milner-Gulland, and T. Clutton-Brock. 2000. The relative roles of density and climatic variation on population dynamics and fecundity rates in three contrasting ungulate species. *Proceedings of the Royal Society of London Series B: Biological Sciences* 267: 1771–1779.

Cransac, M., A. J. R. Hewison, J.-M. Gaillard, et al. 1997. Patterns of mouflon (*Ovis gmelini*) survival under moderate environmental conditions: effects of sex, age, end epizootics. *Canadian Journal of Zoology* 75: 1867–1875.

DelGiudice, G. D., M. R. Riggs, P. Joly, et al. 2002. Winter severity, survival, and cause-specific mortality of female white-tailed deer in north-central Minnesota. *Journal of Wildlife Management* 66: 698–717.

DelGiudice, G. D., J. Fieberg, M. R. Riggs, et al. 2006. A long-term age-specific survival analysis of female white-tailed deer. *Journal of Wildlife Management* 70: 1556–1568.

Eberhardt, L. L. 1977. Optimal policies for conservation of large mammals with special reference to marine ecosystems. *Environmental Conservation* 4: 205–212.

Eberhardt, L. L. 1991. Models of ungulate population dynamics. *Rangifer*, Special Issue 7: 24–29.

Eberhardt, L. L. 2002. A paradigm for population analysis of long-lived vertebrates. *Ecology* 83: 2841–2854.

Ezard, T. H. G., J.-M. Gaillard, M. J. Crawley, and T. Coulson. 2008. Habitat dependence and correlations between elasticities of long-term growth rates. *American Naturalist* 172: 424–430.

Festa-Bianchet, M., T. Coulson, J.-M. Gaillard, et al. 2006. Stochastic predation events and population persistence in bighorn sheep. *Proceedings of the Royal Society B: Biological Sciences* 273: 1537–1543.

Forchhammer, M. C., T. H. Clutton-Brock, J. Lindstrom, et al. 2001. Climate and population density induce long-term cohort variation in a northern ungulate. *Journal of Animal Ecology* 70: 721–729.

Fowler, C. W. 1981. Density dependence as related to life history strategy. *Ecology* 62: 602–610.

Fowler, C. W. 1987. A review of density dependence in populations of large mammals. In *Current Mammalogy*, ed. H. H. Genoways, pp. 401–441. Plenum Press, New York.

Fowler, C. W. 1988. Population-dynamics as related to rate of increase per generation. *Evolutionary Ecology* 2: 197–204.

Fujiwara, M. and H. Caswell. 2002. Estimating population projection matrices from multi-stage mark-recapture data. *Ecology* 83: 3257–3265.

Gaillard, J.-M., J. M. Boutin, D. Delorme, et al. 1997. Early survival in roe deer: causes and consequences of cohort variation in two contrasted populations. *Oecologia* 112: 502–513.

Gaillard, J.-M., D. Delorme, J.-M. Boutin, et al. 1993. Roe deer survival patterns–a comparative analysis of contrasting populations. *Journal of Animal Ecology* 62: 778–791.

Gaillard, J.-M., P. Duncan, S. E. Van Wieren, A. Loison, F. Klein, and D. Maillard. 2008. Managing large herbivores in theory and practice : is the game the same for browsing and grazing species? In *The Ecology of Browsing and Grazing*, Ecological Studies 195, eds. I. J. Gordon and H. H. T. Prins, 293–307. Springer Verlag, Berlin Heidelberg.

Gaillard, J.-M., M. Festa-Bianchet, and N. G. Yoccoz. 1998. Population dynamics of large herbivores: variable recruitment with constant adult survival. *Trends in Ecology and Evolution* 13: 58–63.

Gaillard, J.-M., M. Festa-Bianchet, N. G. Yoccoz, et al. 2000. Temporal variation in fitness components and population dynamics of large herbivores. *Annual Review of Ecology and Systematics* 31: 367–393.

Gaillard, J.-M., A. Loison, C. Toigo, et al. 2003a. Variation in life history traits and realistic population models for wildlife management: the case of ungulates. In *Animal Behavior and Wildlife Conservation*, eds. M. Festa-Bianchet and M. Apollonio, pp. 115–132. Island Press, Washington.

Gaillard, J.-M., A. Loison, C. Toigo, et al. 2003b. Cohort effects and deer population dynamics. *Ecoscience* 10: 412–420.

Gaillard, J.-M., D. Pontier, D. Allainé, J. D. Lebreton, J. Trouvilliez, and J. Clobert. 1989. An analysis of demographic tactics in birds and mammals. *Oikos* 56: 59–79.

Gaillard, J.-M. and N. G. Yoccoz. 2003. Temporal variation in survival of mammals: a case of environmental canalization? *Ecology* 84: 3294–3306.

Garel, M., A. Loison, J.-M. Gaillard, et al. 2004. The effects of a severe drought on mouflon lamb survival. *Proceedings of the Royal Society of London Series B: Biological Sciences* 271: S471–S473.

Garrott, R. A., L. L. Eberhardt, P. J. White, et al. 2003. Climate-induced variation in vital rates of an unharvested large-herbivore population. *Canadian Journal of Zoology* 81: 33–45.

Gilpin, M. E. and F. J. Ayala. 1973. Global models of growth and competition. Proceedings of the National Academy of Sciences of the United States of America 70: 3590–3593.

Grant, A. and T. G. Benton. 2000. Elasticity analysis for density-dependent populations in stochastic environments. *Ecology* 81: 680–693.

Grant, A. and T. G. Benton. 2003. Density-dependent populations require density-dependent elasticity analysis: an illustration using the LPA model of Tribolium. *Journal of Animal Ecology* 72: 94–105.

Hayssen, V. D., and A. Van Tienhoven. 1993. *Asdell's Patterns of Mammalian Reproduction: A Compendium of Species-specific Data*. Comstock Pub. Associates, California.

Hewison, A. J. M., J. P. Vincent, J. M. Angibault, et al. 1999. Tests of estimation of age from tooth wear on roe deer of known age: variation within and among populations. *Canadian Journal of Zoology* 77: 58–67.

Hone, J. and T. H. Clutton-Brock. 2007. Climate, food, density and wildlife population growth rate. *Journal of Animal Ecology* 76: 361–367.

Hosack, D. A., P. S. Miller, J. J. Hervert, et al. 2002. A population viability analysis for the endangered Sonoran pronghorn, *Antilocapra americana sonoriensis*. *Mammalia* 66: 207–229.

Jorgenson, J. T., M. FestaBianchet, J.-M. Gaillard, et al. 1997. Effects of age, sex, disease, and density on survival of bighorn sheep. *Ecology* 78: 1019–1032.

Kjellander, P., J.-M. Gaillard, M. Hewison, et al. 2004. Predation risk and longevity influence variation in fitness of female roe deer (*Capreolus capreolus* L.). *Proceedings of the Royal Society of London Series B: Biological Sciences* 271: S338–S340.

Krebs, C. J. 2002. Two complementary paradigms for analysing population dynamics. *Philosophical Transactions of the Royal Society of London Series B: Biological Sciences* 357: 1211–1219.

Lande, R., S. Engen, and B. E. Saether. 2003. *Stochastic Population Dynamics in Ecology and Conservation*, Oxford Series in Ecology and Evolution. Oxford University Press, Oxford.

Lande, R., S. Engen, B. E. Saether, et al. 2006. Estimating density dependence from time series of population age structure. *American Naturalist* 168: 76–87.

Langvatn, R. and A. Loison. 1999. Consequences of harvesting on age structure, sex ratio and population dynamics of red deer *Cervus elaphus* in Central Norway. *Wildlife Biology* 5: 213–223.

Lebreton, J. D., K. P. Burnham, J. Clobert, et al. 1992. Modeling survival and testing biological hypotheses using marked animals–a unified approach with case-studies. *Ecological Monographs* 62: 67–118.

Lebreton, J. D. and R. Pradel. 2002. Multistate recapture models: modelling incomplete individual histories. *Journal of Applied Statistics* 29: 353–369.

Linnell, J. D. C., R. Aanes, and R. Andersen. 1995. Who killed Bambi? The role of predation in the neonatal mortality of temperate ungulates. *Wildlife Biology* 1: 209–223.

Loison, A., J.-M. Gaillard, and J.-M. Jullien, et al. 1996. Demographic patterns after an epizootic of keratoconjunctivitis in a chamois population. *Journal of Wildlife Management* 60: 517–527.

Loison, A., J.-M. Gaillard, and H. Houssin. 1994. New insight on survivorship of female chamois (*Rupicapra rupicapra*) from observation of marked animals. *Canadian Journal of Zoology* 72: 591–597.

Manly, B. F. J. 1979. Note on key factor-analysis. *Researches on Population Ecology* 21: 30–39.

McLoughlin, P. D., M. S. Boyce, T. Coulson, et al. 2006. Lifetime reproductive success and density-dependent, multi-variable resource selection. *Proceedings of the Royal Society B: Biological Sciences* 273: 1449–1454.

Mills, L. S., D. F. Doak, and M. J. Wisdom. 1999. Reliability of conservation actions based on elasticity analysis of matrix models. *Conservation Biology* 13: 815–829.

Morris, W. F., C. A. Pfister, S. Tuljapurkar, et al. 2008. Longevity can buffer plant and animal populations against changing climatic variability. *Ecology* 89: 19–25.

Nichols, J. D. 1992. Capture-recapture models. *Bioscience* 42: 94–102.

Nichols, J. D., J. R. Sauer, K. H. Pollock, et al. 1992. Estimating transition-probabilities for stage-based population projection matrices using capture recapture data. *Ecology* 73: 306–312.

Nussey, D. H., L. E. B. Kruuk, A. Morris, and T. H. Clutton-Brock. 2007. Environmental conditions in early life influence ageing rates in wild population of red deer. *Current Biology* 17: R1000 -R1001.

Page, R. E. 1988. The inverted pyramid: ecosystem dynamics of wolves and moose on Isle Royale. Dissertation, Michigan Technological University, Houghton, Michigan, USA.

Peacock, E. and D. L. Garshelis. 2006. Comment on "On the regulation of populations of mammals, birds, fish, and insects". IV. *Science* 313: 45.

Pemberton, J. M. 2008. Wild pedigrees: the way forward. *Proceedings of the Royal Society of London Series B: Biological Sciences* 275: 613–621.

Pettorelli, N., J.-M. Gaillard, P. Duncan, et al. 2003. Age and density modify the effects of habitat quality on survival and movements of roe deer. *Ecology* 84: 3307–3316.

Pettorelli, N., J.-M. Gaillard, G. Van Laere, et al. 2002. Variations in adult body mass in roe deer: the effects of population density at birth and of habitat quality. *Proceedings of the Royal Society of London Series B: Biological Sciences* 269: 747–753.

Pettorelli, N., J.-M. Gaillard, N. G. Yoccoz, et al. 2005. The response of fawn survival to changes in habitat quality varies according to cohort quality and spatial scale. *Journal of Animal Ecology* 74: 972–981.

Portier, C., M. Festa-Bianchet, J.-M. Gaillard, et al. 1998. Effects of density and weather on survival of bighorn sheep lambs (*Ovis canadensis*). *Journal of Zoology* 245: 271–278.

Robbins, C. T. and B. L. Robbins. 1979. Fetal and neonatal growth patterns and maternal reproductive effort in ungulates and sub-ungulates; *American Naturalist* 114: 101–116.

Rose, K. E., T. H. Clutton-Brock, and F. E. Guinness. 1998. Cohort variation in male survival and lifetime breeding success in red deer. *Journal of Animal Ecology* 67: 979–986.

Royama, T. 1996. A fundamental problem in key factor analysis. *Ecology* 77: 87–93.

Saether, B. E. and O. Bakke. 2000. Avian life history variation and contribution of demographic traits to the population growth rate. *Ecology* 81: 642–653.

Saether, B. E., S. Engen, E. J. Solberg, et al. 2007. Estimating the growth of a newly established moose population using reproductive value. *Ecography* 30: 417–421.

Saether, B. E. and M. Heim. 1993. Ecological correlates of individual variation in age at maturity in female moose (*Alces alces*)–The effects of environmental variability. *Journal of Animal Ecology* 62: 482–489.

Schröder, W. and I. Elsner-Schack. 1985. Correct age determination in chamois. In *The Biology and Management of Mountain Ungulates*, eds. S. Lovari, pp. 65–70. Croom Helm, London.

Sibly, R. M., D. Barker, M. C. Denham, et al. 2005. On the regulation of populations of mammals, birds, fish, and insects. *Science* 309: 607–610.

Sinclair, A. R. E. 1977. *The African Buffalo: A Study of Resource Limitation of Populations.* University of Chicago Press, Chicago.

Solberg, E. J., P. Jordhoy, O. Strand, et al. 2001. Effects of density-dependence and climate on the dynamics of a Svalbard reindeer population. *Ecography* 24: 441–451.

Solberg, E. J., A. Loison, J.-M. Gaillard, and M. Heim. 2004. Lasting effects of conditions at birth on moose body mass. *Ecography* 27: 677–687.

Stearns, S. C. 1983. The influence of size and phylogeny on patterns of covariation among life-history traits in the mammals. *Oikos* 41: 173–187.

Stearns, S. C. 1992. *The Evolution of Life Histories.* Oxford University Press, New York.

Stubsjoen, T., B. E. Saether, E. J. Solberg, et al. 2000. Moose (*Alces alces*) survival in three populations in northern Norway. *Canadian Journal of Zoology* 78: 1822–1830.

Toïgo, C., J.-M. Gaillard, M. Festa-Bianchet, et al. 2007. Sex- and age-specific survival of the highly dimorphic Alpine ibex: evidence for a conservative life-history tactic. *Journal of Animal Ecology* 76: 679–686.

Tuljapurkar, S. 1989. An uncertain life: Demography in random environments. *Theoretical Population Biology* 35: 227–294.

Vaupel, J. W., K. G. Manton, and E. Stallard. 1979. Impact of heterogeneity in individual frailty on the dynamics of mortality. *Demography* 16: 439–454.

Weladji, R. B., D. R. Klein, O. Holand, and A. Mysterud. 2002. Comparative response of Rangifer tarandus and other northern ungulates to climatic variability. In *Reindeer as a Keystone Species in the North-Biological, Cultural and Socio-economic Aspects*, eds. P. Soppela, W. Ruth, B. Ahman, and J. A. Riseth, Arctic Center Reports, 38, pp. 124–130. University of Lapland, Finland.

Wilson, A. J., J. M. Pemberton, J. G. Pilkington, et al. 2006. Environmental coupling of selection and heritability limits evolution. *PLoS Biology* 4: 1270–1275.

Wittemyer, G., H. B. Rasmussen, and I. Douglas-Hamilton. 2007. Breeding phenology in relation to NDVI variability in free-ranging African elephant. *Ecography* 30: 42–50.

Wittmer, H. U., B. N. McLellan, R. Serrouya, et al. 2007. Changes in landscape composition influence the decline of a threatened woodland caribou population. *Journal of Animal Ecology* 76: 568–579.

Wittmer, H. U., A. R. E. Sinclair, and B. N. McLellan. 2005. The role of predation in the decline and extirpation of woodland caribou. *Oecologia* 144: 257–267.

5

Irruptive dynamics and vegetation interactions

John E. Gross[1], Iain J. Gordon[2] and
Norman Owen-Smith[3]

[1]*Inventory and Monitoring Program, National Park Service, Fort Collins, Colorado, United States of America*

[2]*CSIRO Davies Laboratory, Aitkenvale, Queensland, Australia*

[3]*School of Animal, Plant and Environmental Sciences, University of the Witwatersrand, Johannesburg, South Africa*

Large mammalian herbivores are widely perceived to have a propensity to increase rapidly to high abundance levels, especially in the absence of predation. This situation, an irruption, is generally expected to lead to severe mortality associated with vegetation degradation. The expected long-term degradation of vegetation has been used to justify management interventions to curtail or prevent population growth via some form of culling (Jewell et al. 1981, Gunn et al. 2003). Contributions in McShea et al. (1997) addressed problems of "over-abundance," particularly with respect to chronic high densities of deer in North America. Sinclair (1979) and McCullough (1997) provide historical reviews and an initial conceptual framework for irruptive dynamics. More recently, Forsyth and Caley (2006) reviewed evidence and concepts and evaluated mathematical models that could accommodate the "irruptive paradigm." The features typifying irruptive dynamics include (i) a sustained period of rapid population growth, (ii) changes in vegetation as a direct result of grazing and browsing,

Dynamics of Large Herbivore Populations in Changing Environments, 1st edition. Edited by Norman Owen-Smith.
© 2010 Blackwell Publishing

perceived to be adverse for supporting the population, and (iii) severe mortality reducing the population to a lower abundance level. There could be a single irruptive oscillation, or persistent population instability manifested by repeated increases and crashes in abundance.

The concept of irruptive herbivore dynamics was first articulated by Leopold (1943) to describe patterns exhibited by many deer populations in North America. Leopold drew, in particular, on estimates of the changes in the abundance of mule deer inhabiting the Kaibab plateau in Arizona in the early 1900s, following the elimination of predators. The factual basis of this apparent irruption was discredited by Caughley (1970), but the conceptual model retained its popularity. More recently, Binkley et al. (2006) reexamined Leopold's (1943) claims and presented evidence from tree demography that is consistent with the deer population dynamics Leopold described. Riney (1964) suggested that Leopold's conceptual model applied generally where ungulates were introduced into ecosystems that lacked native large herbivores, in tropical as well as temperate environments. He suggested that the demographic lag associated with delayed recruitment into the reproductive segment contributed to the overshoot in abundance before food shortage caused a population decline. Caughley (1970) used this conceptual framework to analyze his observations of variation in vital rates of Himalayan tahr at increasing distances from the site of their introduction in New Zealand, with supporting but anecdotal observations of severe impacts on key food resources.

In this chapter, we first provide a brief mathematical context that serves to highlight herbivore and vegetation attributes that contribute to irruptive dynamics. These models of irruptive dynamics provide a theoretical framework for comparison with the examples of irruptions that follow. We then focus on identifying circumstances that promote population irruptions from what might be termed more normal dynamics. Is the propensity to irrupt and crash in abundance a feature of particular herbivore species, e.g. those that have a high potential growth rate? Or are irruptive dynamics generated by particular environmental conditions, e.g. restrictions on dispersal coupled with absence of predators? Finally, we note some implications of our insights for management of large herbivores.

5.1   Models of herbivore–vegetation interactions

Models of interactive herbivore–vegetation systems represent the herbivore dynamics as responding to the changing abundance of edible and accessible vegetation, and at the same time reducing the productive capacity of the vegetation (Chapter 2). In these simple models, the vegetation is presumed to constitute a homogeneous resource, and the herbivore population responds to changes in the quantity of edible and accessible food over some multiyear period (e.g. Caughley 1976a). Eventually, herbivore abundance falls to a level allowing vegetation recovery. The reciprocal oscillations in herbivore and vegetation abundance generated may be either dampened, or sustained as a limit cycle, depending on the parameter values. There is in fact a rather narrow range of parameter values that permit persistence of both populations. Greater efficiency in converting forage consumed into herbivore biomass, or a low-maintenance food requirement, leads to a greater overshoot of the sustainable abundance level, and hence wider oscillations. A higher relative vegetation growth rate increases the ability of the vegetation to resist being reduced to low abundance. Slow recovery by the vegetation increases the period and amplitude of the oscillations that are generated.

This model formulation suggests that herbivore populations that have a relatively high potential population growth rate, through early maturation or birth of multiple offspring, have an especially high propensity to exhibit irruptive dynamics. Nevertheless, Caughley (1976b) suggested that this model could also be applied to the dynamics of elephant populations interacting with savanna woodlands. In this application, oscillations developed because of the slow recovery rate of tree populations coupled with efficient feeding by the elephants, despite the slow growth rate of the elephant population. Caughley (1987) examined kangaroo–vegetation interactions by varying vegetation and herbivore feedbacks in the basic model. In the highly variable Australian system he examined (rainfall CV = 0.45), models were run using parameters estimated from extensive field observations. From model projections, Caughley concluded that irruptive dynamics of red kangaroos were driven almost entirely by forage production, with an insignificant role for density-dependent feedbacks.

Turchin (2003) presented a more general treatment of coupled consumer–resource models, and suggested that complex dynamics rather than an approach towards some equilibrium abundance level would usually result from the delays inherent in the interaction between the two populations. He suggested that persistent oscillations were likely to be manifested by large herbivore populations, just as for smaller organisms. However, due to longer response times of large herbivores, the period of the oscillations was likely to be 30 years or greater, and so only apparent in long time series of abundance levels.

Turchin (2003; see also Turchin and Batzli 2001) drew attention to the importance of the vegetation production function in governing the potential for oscillations to be generated. If instead of the logistic function, vegetation regrowth was assumed to be largely independent of the amount of food supporting the herbivores, oscillations would be suppressed. This could be a realistic assumption if much of the vegetation biomass was unavailable to the herbivores, below-ground or too high in the tree canopy. Alternative plant growth equations were also considered by Schmitz and Sinclair (1997).

In applying this modeling approach to the dynamics of a white rhino population interacting with grasses, Owen-Smith (1988) separated the within-year growth dynamics of the grasses, producing food for the herbivore, from between-year changes in the grass population generating this production. Allowance was also made for the effects of rainfall on grass growth and hence on the extent of the over-grazing causing a depression of the grass population. The stability or otherwise of the interaction depended largely on the rate of recovery of the grass population following episodes of over-grazing associated with low rainfall years.

Two factors, absent from the models above, that may confer stability are a gradual reduction in forage quality with consumption, and broader-scale ecosystem heterogeneity. Conversely, Ellis and Swift (1988) argued that high variance in rainfall (CV > 30%), and hence food resources, led to intrinsically disequilibrial dynamics of ungulates in semi-arid rangelands. Others have interpreted large fluctuations of herbivore numbers in semi-arid systems as strongly density-dependent responses, linked to the availability of key resources during the dry season and drought years (Illius and O'Connor 1999, 2000). These studies, focused on domestic livestock, offer competing hypotheses on the long-term impacts of high herbivore abundances on the vegetation.

Owen-Smith (2002a,b) used a metaphysiological modeling approach (Getz 1993) to evaluate the effects of heterogeneity in the vegetation resource on the population stability of large herbivores. This model accommodated seasonal variation in vegetation growth as well as the adaptive responses of the herbivores in their diet selection to changes in the availability of different food types. Owen-Smith (2002a,b) constructed a model that simulated ungulate biomass, forage growth and consumption, and spatial heterogeneity via multiple patch types that offered forages that differed in quality. Patches with low-quality forage stabilized the herbivore population dynamics by providing a "reserve" when forages in high-quality patches were depleted, and through buffering against extreme food depletion. When the forage available was mostly of high nutritional value, the ungulate population grew until little vegetation remained, and then crashed.

Forsyth and Caley (2006) compared the θ-logistic and time-delayed logistic models, plus two new models, in terms of their success in replicating observed cases of irruptive dynamics. Rather than representing the dynamics of forage resources explicitly, two of the models incorporated a linear

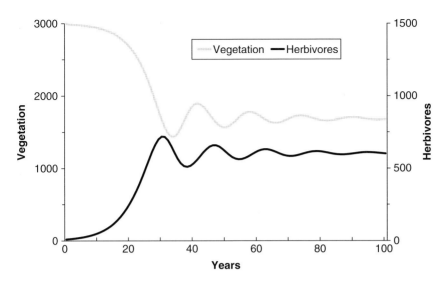

Figure 5.1 Output of Caughley's (1976a) model suggesting the irruptive trajectory with dampened oscillations potentially shown by an introduced large herbivore population interacting with vegetation.

decline in "surplus" resources from the time of initial colonization until all of this surplus was consumed. One new model included a variable time lag. Each of the four models was most strongly supported by at least one data set, although models with resource depletion were clearly most suitable for populations that exhibited the classical "growth and crash" irruptive dynamics (Fig. 5.1). Forsyth and Caley's (2006) models emphasized the potential contribution of lag times and resource depletion to generating irruptive dynamics. They found support for models incorporating lag times of 3 to 12 years for five of the seven ungulate populations they evaluated. These results are broadly consistent with some other observations. For example, white-tailed deer recruitment rates and male body mass appeared most strongly related to density with a 3-year lag (Fryxell et al. 1991). Forsyth and Caley (2006) noted the importance of examining models structurally capable of representing irruptive dynamics, such as the new models they offered. However, the weak mechanistic basis of the models limits their assessment of ecological factors that promote irruptions.

Time lags have traditionally been attributed to vegetation depletion by herbivores and the time necessary for vegetation recovery. However, delayed density dependence can also result from cohort effects, with individuals born under high-density conditions exhibiting reduced survival and reproductive performance compared to those from the population during the phase of high population growth (red deer, Albon et al. 1987; reindeer, Crete and Huot 1993; white-tailed deer; Fryxell et al. 1991).

Mathematical models of irruptions have focused on time lags and resource depletion. The relative importance of changes in forage quality and distribution, and the roles of predators and density-independent effects, such as extreme weather, have rarely been considered.

5.2 Examples of irruptive dynamics

5.2.1 Island populations

Populations of ungulates on islands seem particularly prone to extreme irruptions. Examples include reindeer introduced to St Matthews Island in the Pribiloffs, which increased from a nucleus of 29 animals in 1944 to a maximum of about 6000 in 1963, and thereafter crashed over winter to fewer than 50 animals by the following summer (Klein 1968). The extreme mortality was originally ascribed to the almost complete elimination of lichen from the island, but Miller et al. (2005) showed that the die-off also coincided with extreme winter conditions in the form of icing of the snow cover. McCullough (1997) described and discussed the irruptive dynamics of deer introduced into two islands (North Manitou, MI; Angel Island, CA), and the fenced-in George Reserve (MI). Island characteristics thought to promote irruptive dynamics include an absence of predators, a limited and confined area, and relatively homogeneous habitats (Woodgerd 1964, Klein 1968, McCullough 1997). In these circumstances, an ungulate population can grow to population levels at which most available forage is consumed over the course of a year. In the absence of adequate nutrition, the population may be highly susceptible to a crash precipitated by disease, extreme cold weather, or lack of food.

Irruptive dynamics similar to those of white-tailed deer populations have been exhibited by sika deer introduced onto islands or enclosed areas in Japan. For example, on Cape Shiretoko the deer population increased to a peak density of 118 animals per km^2 within the 5 km^2 extent, at a growth rate of 19% per year (Kaji et al. 2004). Severe mortality then reduced the population by two-thirds following a severe winter, but 3 years later the population had almost regained its peak abundance (Fig. 5.2).

Moose on Isle Royale, a 544 km^2 island in Lake Superior (on the United States–Canada border), have exhibited one of the most dramatic and well-documented examples of repeated population irruptions (Fig. 1.10). Moose on Isle Royale are thought to have rapidly increased until about 1935, when there was apparently a crash to a much lower density. Between the mid-1800s and the late 1900s, moose browsing was apparently responsible

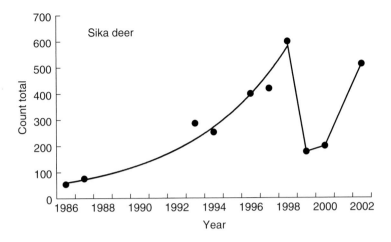

Figure 5.2 Irruptive trajectory shown by Sika deer in the Cape Shiretoko region (5 km²) of Shiretoko National Park in Japan (from Kaji et al. 2004).

for reducing the abundance of their primary browse, balsam fir, from about half the canopy to less than 5% (McLaren and Peterson 1994). This dramatic reduction in balsam fir was not observed on nearby islands uninhabited by moose, and patterns of balsam fir growth on Isle Royale did not respond to climate fluctuations as did fir in other areas. This reduction of preferred browse did not prevent a second irruption towards a peak density exceeding 4 moose per km² (Post et al. 2002).

Moose populations from other areas are not known to exhibit such dramatic population irruptions or cycles despite the high population-growth rate enabled by their capacity for twinning. The Isle Royale population stands out for its cyclic behavior (Messier and Crete 1985). Moose on Isle Royale occur at an unusually high density (Peterson 1999), and the manifestation of irruptive population patterns on Isle Royale is consistent with populations of other species inhabiting islands or confined areas (Klein 1968, McCullough 1997). The wolf population crash on Isle Royale may have precipitated the irruption of moose, and wolf predation appears to be a strong determinant of average or equilibrium moose population size, although predation has a relatively small influence on interannual variation about the mean size (Vucetich and Peterson 2004a).

Soay sheep on the Scottish island of Hirta have shown repeated increases followed by crashes in abundance, with no reduction in the peak abundance levels attained (Fig. 1.4). The unusual feature is the short duration of the increase phase, lasting only 2–3 years. Population crashes are generally associated with adverse weather in late winter and are predisposed by a population structure containing a high proportion of vulnerable young animals and old males. Population growth during the increase phase has averaged over 30% per year, while the magnitude of single-year declines has averaged nearly 50% (Coulson et al. 2008). The lack of any reduction in the productive capacity of the vegetation is perhaps a result of high rainfall and highly fertile soils.

5.2.2   Mainland populations

Within North America, bison, elk, deer, and moose populations achieved high densities only when they coexisted with a single species of predator (Peterson et al. 2003). A single species of predator was typically incapable of strong population regulation, while intact predator communities may suppress irruptive oscillations by slowing the rate of prey population growth during the increase phase (Wilmers et al. 2007). Regardless of the ability of predators to inhibit or prevent a rapid population increase, a lagged numerical response by predators will amplify or enhance the rate of decline following a peak, and predation may have a strong effect during a recovery phase (e.g., Turchin 2003).

Ibex introduced into the Swiss National Park in the absence of predators exemplify the classical oscillatory pattern projected by models (Fig. 5.3).

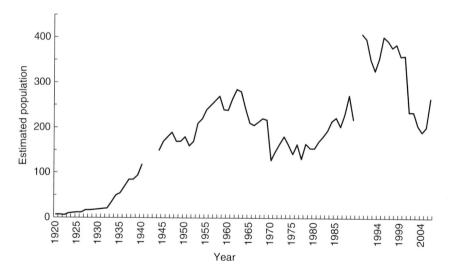

Figure 5.3 Irruptive dynamics of ibex introduced in the Swiss National Park. Note that there was a change in census method in 1990 yielding a higher estimated total (data from F. Filli, personal communication 2008).

The long-term dynamics of the pronghorn population within the Greater Yellowstone ecosystem have also been interpreted as conforming to an irruptive pattern (White et al. 2007). The population grew from about 300 in 1920 to more than 800 animals in 1938. A high population was maintained for about a decade, when culling was instituted in response to apparent degradation of the sagebrush which formed their winter food resource (Houston 1982). The population then crashed to fewer than 200 animals, but had recovered to a total of 600 by 1991, when severe mortality reduced the population by almost two-thirds. Subsequently, the population remained at this lower abundance of around 200 animals. Interpretation of these results is confounded by increases in the abundance of elk and mule deer, which may have affected the availability of sagebrush. The contribution by sagebrush to the winter diet of the pronghorn decreased from 70% during 1985–8 to less than 10% by 2001, being replaced by less nutritious rabbit brush (Singer and Norland 1994, Boccadori 2002).

Some of the best examples of repeated phases of growth in abundance followed by sharp population declines come from studies of migratory tundra caribou in Canada, Alaska, and Greenland. Their northern habitats are beyond the extent of agriculture and widespread successional vegetation changes following logging. Most populations are hunted in Alaska and Canada (but not Greenland) and are subject to predation by both grizzly bear and wolves. Despite being in a two-predator (but one main prey) system, the herds have fluctuated in abundance over an 11-fold range on average (extremes 3–24 fold), with the interval between peaks generally being 40–70 years (reviewed by Gunn 2003). Some degree of regional synchrony in trends among caribou herds is evident in eastern Canada and Greenland, seemingly linked to climatic variation associated with the North Atlantic Oscillation (NAO; Forschhammer et al. 2002). In western North America, under the influence of the North Pacific Oscillation, synchrony of regular fluctuations among adjacent herds appears less entrained as the phase of low numbers is more variable (Fig. 5.4). Relationships among the timing and extent of forage production, growth in body mass, pregnancy rates, and offspring survival have been demonstrated, indicating an underlying influence through resource availability (Griffith et al. 2002).

In North American herds, the shift from increasing to decreasing trend can be abrupt, and apparently results from depletion of vegetation resources on the summer range rather than of lichens in the winter range. The consequent deterioration in body condition of adult females during this critical period leads to a reduction in growth and survival of offspring (Couturier et al. 1990, Crête and Huot 1993). Food availability during winter seems to be generally less influential. The effect of food limitation

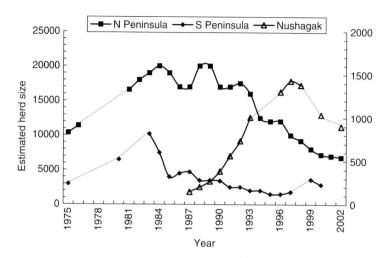

Figure 5.4 Fluctuations in abundance shown by three caribou herds in south-western Alaska suggesting different stages of an irruptive oscillation with a period of 40–50 years (from Valkenburg et al. 2003).

during early ontogeny on the body mass of the juveniles can be persistent, giving rise to variation among cohorts in their later reproductive success and susceptibility to predation. Although predation may slow the rate of population growth or accelerate the rate of decrease, it otherwise does not seem to affect the general pattern of the dynamics exhibited by the wide-ranging herds.

5.2.3 African examples

The huge increase in abundance of wildebeest in the Serengeti ecosystem has been presented as an irruptive increase following the elimination of rinderpest. However, despite the high population density attained (around 60 animals/km^2; Mduma et al. 1999), the wildebeest population has maintained this abundance apart from temporary decline by 30% associated with the severe drought in 1993–4. Predation has little effect on the population of these migrants, and migration may prevent local overgrazing from developing. Furthermore, the Serengeti grasslands have a high productive capacity and have co-evolved with an exceptionally diverse and dense community of herbivores.

Considering the diversity of large African herbivores, there are relatively few examples of irruptive dynamics by native African ungulate populations, perhaps because populations in small reserves are generally subject to culling to prevent high abundance levels from developing. One case is that of waterbuck, which attained a density of around 40 animals per km^2 in the early 1970s, before crashing to under 10 animals per km^2, in the fenced Nakuru National Park in Kenya (extent 140 km^2) (Mwangi 1998). Gazelle numbers also declined drastically around the same time, but species that had initially been at low density showed increasing trends. Severe population crashes following attainment of high density were documented for reedbuck in Kyle National Park, Zimbabwe (Ferrar and Kerr 1971) and nyala in Ndumu Game Reserve, South Africa (Keep 1973). In both cases, the heavy mortality was associated with cold and wet weather. Extreme mortality, reducing populations of several ungulates by 80–90% in the Klaserie Private Nature Reserve in South Africa, occurred during a severe drought, but followed the buildup in abundance levels of wildebeest and other grazers to about four times that in the nearby Kruger National Park as a result of the widespread provision of waterpoints. In the Kruger National Park, cyclic variation in the abundance of several ungulate populations over a more than two-fold range was related to cyclic variation in rainfall conditions, but driven more by shifting vulnerability to predation rather than changing food supplies (Ogutu and Owen-Smith 2005). Further examples where large herbivore populations have crashed by 50% or more are provided by Erb and Boyce (1999) and Reed et al. (2003).

5.3   Effects of irruptions on vegetation

The conceptual model of irruptive dynamics assumes that a high density of ungulates depresses food resources to the detriment of the capacity of the vegetation to support the herbivore population (Leopold 1943, Caughley 1970). There is ample evidence that grazing and browsing impacts from large herbivores can substantially alter vegetation cover and composition (Côte et al. 2004, McShea 2005, Mysterud 2006, Ward 2006). While grazing can have profound impacts on vegetation, Milchunas and Lauenroth (1993) found that grazing pressure ranked third in accounting for vegetation change, behind evolutionary history and aboveground net primary production, in short-grass steppe in North America and subhumid African grasslands.

   Hansen et al. (2007) documented the almost complete elimination of fruiticose lichen by reindeer introduced into a previously ungrazed valley in Svalbard, plus reductions in cover of mosses and vascular plants. This led to a crash of the reindeer population from 360 to 80 animals when icing of the snow cover blocked access to forage (Aanes et al. 2000). Mosses recovered and rapidly exceeded their initial abundance. Among vascular plants, graminoids increased while other important forage species remained depressed and lichens did not recover. Year-round grazing by these resident reindeer may have inhibited vegetation recovery following the reduction in the reindeer population (Hansen et al. 2007).

In other situations, herbivores have persisted at high abundance despite dramatic vegetation change. This includes the white-tailed deer population in Anticosta Island (7943 km^2) in Quebec, where deer at a density of 20 animals per km^2 have greatly reduced the abundance of preferred deciduous browse and balsam fir (Tremblay et al. 2005). Litter-fall now provides much of the food consumed by these deer during winter, with the production of this reserve resource uninfluenced by the deer abundance. The Anticosta deer population has maintained a high population density despite substantial reduction in diet quality (Simard et al. 2008) and a continual reduction in preferred forages since at least the 1950s (Tremblay et al. 2005). Elk in Yellowstone National Park have largely maintained a high abundance level despite the extreme reduction in favored vegetation components such as aspen and willow (although these preferred browse species occupy only a small proportion of the landscape) (NRC 2002). Ibex introduced into Gran Paradiso National Park in Italy (Jacobson et al. 2004) and Belledonne in France (Toigo et al. 2007) have reached a stable population ceiling without evidence of vegetation degradation.

In many situations, persistently heavy grazing or browsing has led to the replacement of preferred plant species by other species that are less nutritious and thus less able to support the herbivore population (Mott 1987; chapters in Danell et al. 2006). In other cases, grazing-tolerant species that are more nutritious have taken over (e.g. grazing lawns, McNaughton 1984). Such vegetation changes may be facilitated by increased nutrient cycling enabled by dung and urine contributions from the herbivores (Hobbs et al. 1991, McNaughton et al. 1997, Pastor et al. 2006). However, selective herbivory concentrated on nitrogen-fixing plants or those with rapid litter decomposition can also have negative effects on nutrient cycling (Pastor and Cohen 1997, Ritchie et al. 1998).

5.4   Changing perspectives

From the examples above, we identified five factors consistently associated with irruptive dynamics of ungulates: (i) initial "surplus" forage, (ii) vegetation changes, (iii) lack of effective predation, (iv) confined populations subject to restrictions on movements, and (v) extreme weather. The relative influences of these factors in promoting irruptive features has not previously been adequately explored.

5.4.1   Surplus forage

Leopold (1943) emphasized the central role of "surplus" forage in fueling population growth to densities well above those that could be supported by annual forage production. However, early concepts vary considerably from our current understanding. The traditional view assumed that preferred forages were rapidly consumed during an irruption, leading to structural changes in vegetation (e.g., browse lines), and the eventual replacement of preferred forages with less-preferred species. This model does not explicitly link population dynamics to ecosystem productivity or standing biomass. The contributions of a forage "surplus" are most apparent in low-productivity arctic environments, where the accumulated biomass, rather than annual forage production, fuels the growth of herbivore populations to densities well above sustainable levels. As an example, reindeer on St Paul Island (Pribilof Islands, Alaska) decimated the abundant lichen flora, an important winter food source, after their introduction in 1911 (Scheffer 1951). The St Paul Island reindeer population achieved a peak population of about 2000 in 1939, and the herd then rapidly declined to fewer than 300 animals by 1946 (Scheffer 1951). Scheffer attributed the decline, in part, to four consecutive harsh winters, including one with severe icing, and in large part to nearly total consumption of the lichen mat. In Svalbard, lichens were almost completely replaced by mosses, a much less nutritious forage, following the introduction of reindeer (Hansen et al. 2007). On Isle Royale, the irruption by moose was promoted by the initial abundance of balsam fir, a preferred winter forage. Balsam fir declined in cover from about 46% prior to the arrival of moose to less than 5% after moose had inhabited the island for eight decades (McLaren and Peterson 1994). A detailed analysis suggested that balsam fir had an important role in determining the moose dynamics (Vucetich and Peterson 2004a).

The concept of "surplus" forage is important because it links accumulated biomass, above and beyond annual vegetation growth, to the likelihood and magnitude of an irruption. Following this logic, ungulate populations inhabiting grasslands should be less prone to irruptions, in part because grasslands generally do not accumulate a large, persistent mass of high-quality forage (but "decreaser" grasses eliminated by intense grazing could play this role) because grass quality declines as it matures. In contrast, lichen-dominated arctic winter ranges may accumulate a deep lichen mat from primary production over many years and consumption of this "surplus" forage can support high rates of overwinter survival by resident reindeer (Scheffer 1951, Klein 1968, Aanes et al. 2000).

5.4.2   Vegetation changes

Models of herbivore irruptions clearly identified the rate of forage regeneration relative to ungulate population growth as a key determinant of the magnitude of post-peak crashes of irruptive populations, beginning with Caughley (1976a) and continuing through Turchin (2003). The lag in the resource-dependent feedback restricting herbivore population growth can theoretically lead to a progressive depletion of forage, precipitating severe herbivore mortality through starvation. However, in most real-world situations, the effects of vegetation changes cannot be disentangled from the effects of weather. Nonetheless, foraging by irruptive ungulate populations has led to significant changes in vegetation composition in arctic (Klein 1968, Aanes et al. 2000, Hansen et al. 2007), temperate (McLaren and Peterson 1994), and semiarid (Binkley et al. 2006) environments. Considerable variation exists in forage quality, accessibility, and regeneration potential among vegetation components, and increases in less-nutritious components could play a buffering role by restricting the extent of starvation or malnutrition (Chapter 6). On the other hand, a shift towards more nutritious plant types could make populations more susceptible to oscillations through more extreme depletion of food resources.

5.4.3   Predators

Leopold (1943) asserted that the lack of predators was the most important factor promoting ungulate irruptions. While the presence of wolves on Isle Royale has not prevented the moose population from showing irruptive dynamics, these cursorial predators concentrate their kills mainly on very young or very old moose (Vucetich and Peterson 2004b), meaning that their population impact is reduced compared with ambush predators imposing additive predation even on animals of prime age (Wilmers et al. 2007). Lags in the predator response to changing prey abundance can also promote oscillations in herbivore abundance, and the challenge for cyclic small mammal populations has been to separate the herbivore–predator from the herbivore–vegetation interaction (Turchin 2003). Ineffective or complete lack of predation is associated with many irruptions of confined populations (e.g., Scheffer 1951, Klein 1968, McCullough 1972, 1997, Kaji et al. 2004, Coulson et al. 2008) and is one of the most common factors associated with irruptive dynamics.

A continuing challenge is to evaluate the influence of predation on irruptive dynamics as it interacts with weather, forage availability, and other processes. A lagged numerical response by predators can generate oscillations (e.g., Turchin 2003). If extreme weather causes greater rates of mortality in large herbivores than their predators – either via direct effects on foods (e.g., Putkonen and Roe 2003) or by making prey more subject to predation – then the intuitive result is that predation will amplify the magnitude of population fluctuations. However, this model is not supported by observations on moose in Isle Royale (Vucetich and Peterson 2004a).

5.4.4   Restricted movements

A key consequence of confinement, via water, fences, or other topographical barriers, is to restrict the spatial domain of the system and thereby reduce access to variation in forage and other resources. One challenge is thus to establish which aspects of ecosystem heterogeneity are functionally important in shaping the population dynamics of ungulates (Chapter 6, this volume). Heterogeneity in food quality can potentially buffer against temporal variability by essentially providing forage "reserves" that may restrict rates of starvation (Ellis and Swift 1988, Illius and O'Connor 2000, Owen-Smith 2002a; Chapter 6). Fowler (1987) postulated that a common density-dependent response in large mammal populations is increased rates of emigration, an option obviously not available to most island or fenced populations. The ability to emigrate or exploit spatial heterogeneity is critical to population stability if the density feedback is primarily indirect, via the impact on food resources, with consequent delays. Caughley and Krebs (1983) asserted that there was very little evidence for direct intrinsic regulation among large herbivores, a conclusion supported by Fowler's (1987) more comprehensive review. Body size-related modes of population regulation lead to the hypothesis that constraints on movement have fundamentally different consequences for large versus small herbivores, including effects on the tendency of the population to exhibit irruptive dynamics.

5.4.5   Weather extremes

Harsh weather commonly precipitates population crashes because of the interaction between weather and prior food availability (Chapter 3). Populations that have escaped predation and become limited directly by food resources are vulnerable to variable climatic conditions that may result in low forage production (droughts; Caughley et al. 1985), less food remaining available during critical seasonal periods (ice crusting; Putkonen and Roe 2003), or elevated metabolic demands from cold and wet weather (Keep 1973, Owen-Smith 2000). Climate variability may interact with or disrupt intrinsic propensities towards irruptive dynamics, leading to an irregular period or a more complex pattern of variation in abundance (Ogutu and Owen-Smith 2005).

There are many examples where unusually harsh weather resulted in major population die-offs or crashes (see Chapter 3). More interesting and unresolved questions concern the intrinsic dynamics of large herbivores inhabiting systems with highly variable and unpredictable weather. Caughley and Gunn (2003) compared dynamics of kangaroos and caribou, emphasizing the simple structure of the desert and arctic systems, and the strikingly similar effects of weather in reducing or promoting forage availability. Both herbivore species typically exhibit wide fluctuations in abundance associated with weather. Analyses of a simple model of the kangaroo systems showed that population trends (up or down), on a scale of one or two decades, were intrinsic to the system and driven by aperiodic fluctuations in weather. Similar mechanisms may also account for observations of caribou.

5.5   Synthesis

Here we articulate our conceptual model for irruptions. It is best considered as a set of hypotheses; some of the hypotheses are strongly grounded in observations, while others are yet to be critically examined.

Across the entire spectrum of ungulates, relatively few species have actually exhibited the full irruptive cycle of increase, peak, and crash. McShea (2005) noted that only a small proportion of cervid populations are reported to be "overabundant." Most documented irruptions have been from populations confined by topography or fences. Irruptions may be more common in places where the majority of plants have not been adapted to mammalian grazing or browsing, as in New Zealand or on islands previously free of big herbivores.

Very wide fluctuations in weather, as in Australia, the more arid parts of Africa, and at high latitudes, can lead to essentially disequilibrial dynamics (Walker et al. 1987, Ellis and Swift 1988, Caughley and Gunn 2003). The prevalence of irruptive dynamics in reindeer and caribou populations may be a consequence of the slow rate of recovery of their main winter food resource in cold, nutrient-poor arctic environments. Variation in weather can thus interrupt what might otherwise be a more regular cycle of herbivore–plant dynamics. Alternatively, in some systems, highly variable weather may entrain multi-decadal population trends, with crashes typically precipitated by extreme weather, potentially allowing some vegetation recovery before the herbivore population regains high abundance (e.g., Caughley 1987, Caughley and Gunn 2003).

Irruptive dynamics have often been associated with the absence of significant predation, allowing the herbivore population to grow rapidly towards the ceiling level where food resources become directly limiting. Besides the direct starvation that may result during periods of adverse weather conditions, animals born under high-density conditions may show restricted growth and hence lowered fitness overall, making them less robust to environmental adversity (Gaillard et al., Chapter 4). Questions arise as to the extent to which megaherbivores like elephants and rhinos, essentially free of predation except on juveniles, might be prone to irruptive dynamics because of their huge impact on vegetation, despite slow population growth (compare Caughley 1976b, Owen-Smith 1988, Duffy et al. 1999).

Populations on islands experience not only restricted dispersal, but a reduction in the environmental heterogeneity that might have provided a buffer against die-offs. Die-offs rarely result simply from overabundance and consequent impacts on food resources. The preferred resources are depleted first, slowing herbivore population growth, with these plants being replaced by others of lower quality or more resistant to herbivory. The severity of die-offs depends on the effective environmental heterogeneity as provided through resource buffers as well as the recovery rate of the key plant resources. There is commonly no indication of lowered ceiling levels in successive irruptive peaks, contrary to Caughley's (1976a) model, because populations do not have time to attain the huge abundance that could be supported if environments remained constantly favorable before adverse weather intervenes (e.g. Soay sheep, Chapter 1; deer, McCullough 1997). In temperate or tropical environments, fluctuations in abundance in response to environment variation (e.g., droughts) are likely to be wider in fertile regions where plants are mostly nutritious and hence mostly eaten, compared with nutrient-poor areas where unpalatable plants constitute a buffer (Owen-Smith 2002a). In arid environments, where virtually all nutritious forage is consumed during extended droughts, soil fertility may be much less relevant.

5.6   Implications for conservation and management

Decisions on management of large herbivores hinge on the philosophical basis of management. A decision to cull animals to prevent ecosystem damage is founded on a desire to maintain a particular state, and a belief that it is indeed possible to do so via active management – the "command and control" strategy described by Holling and Meffe (1996). Alternatively, managing for natural ecological processes accommodates and embraces variation, including wide swings in the abundance of plants and animals resulting from what may be intrinsically determined irruptive dynamics.

Historically, many wildlife managers felt that the appropriate management response to the threat of an irruption was to curtail herbivore population growth and prevent "ecosystem damage." This belief had profound implications for management of wildlife. For example, the desire to prevent "ecosystem damage" led the United States National Park Service (NPS) to intensively cull elk and bison in Yellowstone National Park, maintaining the populations of these species below about 12,000 and 1500 animals, respectively, until 1967 (Yellowstone National Park 1997). In 1967, the NPS adopted a policy of "natural regulation" and culling of elk and bison populations in Yellowstone National Park ceased, permitting rapid population increases consistent with the increase stage of an irruption (NRC 2002, figures 4-2 and 4-4; Fig. 1.16 in this book). Neither elk nor bison have exhibited dramatic population crashes, perhaps because animals that left the park were shot by hunters or culled for disease management. A recent evaluation of Yellowstone's northern range concluded that post-1967 populations – of the order of 16,000 elk and 3000 bison – were not impacting the park's ecosystems in ways that justified intensive population control within the park (NRC 2002), although the health of Yellowstone's vegetation remains controversial (Wagner 2006).

Model results support the existence of intrinsic cycling by herbivores, vegetation, and predator communities (e.g., Caughley 1987, Turchin 2003), suggesting that traditional "command and control" management is inappropriate where the objective is to preserve nature in an unimpaired condition. It may be practically impossible to suppress natural variation in population size in some highly variable and extensive systems. In the context of rapidly changing climate and other global forces, management of natural areas to maximize ecosystem resilience and preserve natural fluctuations in herbivore populations is a more laudable and more likely attainable goal than attempting to sustain any particular, perhaps transitory, ecosystem state. Furthermore, evidence is accumulating that long-term impacts or "ecosystem degradation" by irruptive herbivores inhabiting extensive landscapes is the exception rather than the rule. Gunn et al. (2003, 2006) reviewed reindeer crashes and subsequent recovery, concluding that active management may generally be unnecessary for unconfined populations. While we support policies that focus on natural ecological processes, we also acknowledge the importance of heterogeneity and the issue of spatial scale.

When large herbivores are largely restricted to protected areas, it is much more difficult to reconcile desires to preserve natural processes and simultaneously avoid impacts to soil, plants, and other resources. The absence of migration routes, effective predators, and intrinsic mechanisms for population regulation predisposes large herbivores to achieve high densities in locations where they otherwise may not. The five common factors we identified earlier that are associated with irruptions provide a working hypothesis to support management decisions: large herbivores inhabiting systems where these conditions naturally prevail will generally be expected to exhibit wide fluctuations in abundance, and management plans should thus seriously consider the role of naturally occurring wide fluctuations in abundance. Conversely, where the opposite conditions prevail, we would generally expect more stable populations. A key unresolved question concerns the spatial scale of "intact" plant–herbivore systems. Our review and insights from Chapter 6 emphasize the importance of heterogeneity, but we still have a limited ability to inform managers whether the area occupied by a population is "large enough" to dampen or support intrinsic dynamics under a policy of natural regulation.

Acknowledgments

We thank John Fryxell, Jason Marshal, P. J. White, and, especially, Anne Gunn for reviewing earlier drafts of the manuscript and providing suggestions that greatly improved the chapter.

References

Aanes, R., B. E. Saether, and N. A. Oritsland. 2000. Fluctuations of an introduced population of Svalbard reindeer: the effects of density dependence and climatic variation. *Ecography* 23: 437–443.

Albon, S. D., T. H. Clutton-Brock, and F. E. Guiness. 1987. Early development and population dynamics of red deer. II. Density-independent effects and cohort variation. *Journal of Animal Ecology* 56: 69–81.

Binkley, D., M. M. Moore, W. H. Romme, and P. M. Brown. 2006. Was Aldo Leopold right about the Kaibab deer herd? *Ecosystems* 9: 227–241.

Boccadori, S. J. 2002. Effects of Winter Range on a Pronghorn Population in Yellowstone National Park, Wyoming. Thesis, Montana State University, Bozeman, Montana, USA.

Caughley, G. 1970. Eruption of ungulate populations, with emphasis on Himalayan thar in New Zealand. *Ecology* 51: 53–71.

Caughley, G. 1976a. Plant-herbivore systems. In *Theoretical Ecology*, ed. R. M. May, pp. 94–113. Blackwell, London.

Caughley, G. 1976b. The elephant problem–an alternative hypothesis. *East African Wildlife Journal* 14: 265–283.

Caughley, G., G. C. Grigg, and L. Smith. 1985. The effect of drought on kangaroo populations. *Journal of Wildlife Management* 49: 679–685.

Caughley, G. 1987. Ecological relationships. In *Kangaroos: Their Ecology and Management in the Sheep Rangelands of Australia*, eds. G. Caughley, N. Shepherd, and J. Short, pp. 159–187. Cambridge University Press, Cambridge.

Caughley, G. and A. Gunn. 2003. Dynamics of large herbivores in deserts–kangaroos and caribou. *Oikos* 67: 47–55.

Caughley, G. and C. J. Krebs. 1983. Are big mammals simply little mammals writ large? *Oecologia* 59: 7–17.

Côte, S. D., T. P. Rooney, J. P. Tremblay, C. Dussault, and D. M. Waller. 2004. Ecological impacts of deer overabundance. *Annual Review of Ecology Evolution and Systematics* 35: 113–147.

Coulson, T., T. H. G. Ezard, F. Pelletier, et al. 2008. Estimating the functional form of the density dependence from life history data. *Ecology* 89: 1661–1674.

Couturier, S., J. Brunellle, D. Vandal, and G. St Martin. 1990. Changes in the population dynamics of the George River caribou herd, 1976–1987. *Arctic* 43: 9–20.

Crete, M. and J. Huot. 1993. Regulation of a large herd of migratory caribou–summer nutrition affects calf growth and body reserves of dams. *Canadian Journal of Zoology* 71: 2291–2296.

Danell, K., R. Bergström, P. Duncan, J. Pastor, and H. Olff. 2006. *Large Herbivore Ecology and Ecosystem Dynamics*. Cambridge University Press, Cambridge.

Duffy, K. J., B. R. Page, J. H. Swart, and V. B. Bajic. 1999. Realistic parameter assessment for a well known elephant-tree ecosystem model reveals that limit cycles are unlikely. *Ecological Modelling* 121: 115–125.

Ellis, J. E. and D. M. Swift. 1988. Stability of African pastoral ecosystems: alternate paradigms and implications for development. *Journal of Range Management* 41: 450–459.

Erb, J. D. and M. S. Boyce. 1999. Distribution of population declines in large mammals. *Conservation Biology* 13: 199–201.

Ferrar, A. A. and M. A. Kerr. 1971. A population crash of the reedbuck in Kyle National Park, Rhodesia. *Arnoldia (Rhodesia)* 5 (16): 1–19.

Forchhammer, M., E. Post, N. C. Stenseth, and D. M. Boertmann. 2002. Long-term responses in arctic ungulate dynamics to changes in climate and trophic processes. *Population Ecology* 44: 113–120.

Forsyth, D. M. and P. Caley. 2006. Testing the irruptive paradigm of large herbivore dynamics. *Ecology* 87: 297–303.

Fowler, C. W. 1987. A review of density dependence in populations of large mammals. In *Current Mammalogy*, ed. H. H. Genoways, pp. 401–441. Plenum Press, New York.

Fryxell, J. M., D. J. T. Hussell, A. B. Lambert, and P. C. Smith. 1991. Time lags and population fluctuations in white-tailed deer. *Journal of Wildlife Management* 55: 377–385.

Getz, W. M. 1993. Metaphysiological and evolutionary dynamics of populations exploiting constant and interactive resources: r-K selection revisited. *Evolutionary Ecology* 7: 287–305.

Griffith, B., D. C. Douglas, N. E. Walsh, et al. 2002. The Porcupine caribou herd. In *Arctic Refuge Coastal Plain Terrestrial Wildlife Research Summaries*, eds. D. C. Douglas, P. E. Reynolds, and E. B. Rhode, USGS Biological Science Report, pp. 8–37. US Geological Survey, USGS/BRD/BSR-2002-0001.

Gunn, A. 2003. Voles, lemmings and caribou - population cycles revisited? *Rangifer* Special Issue 14: 105–111.

Gunn, A., F. L. Miller, and S. J. Barry. 2003. Conservation of erupting ungulate populations on islands–a comment. *Rangifer* 24: 3–12.

Gunn, A., F. L. Miller, S. J. Barry, and A. Buchen. 2006. A near-total decline in caribou on Prince of Wales, Somerset, and Russell Islands, Canadian Arctic. *Arctic* 59: 1–13.

Hansen, B. B., S. Henriksen, R. Aanes, and B. E. Saether. 2007. Ungulate impact on vegetation in a two-level trophic system. *Polar Biology* 30: 549–558.

Hobbs, N. T., D. S. Schimel, C. E. Owensby, and D. J. Ojima. 1991. Fire and grazing in the tallgrass prairie: contingent effects on nitrogen budgets. *Ecology* 72: 1374–1382.

Holling, C. S. and G. K. Meffe. 1996. Command and control and the pathology of natural resource management. *Conservation Biology* 10: 328–337.

Houston, D. B. 1982. *The Northern Yellowstone Elk: Ecology and Management*. MacMillan Publishing Company, New York.

Illius, A. W. and T. G. O'Connor. 1999. On the relevance of nonequilibrium concepts to arid and semiarid grazing systems. *Ecological Applications* 9: 798–813.

Illius, A. W. and T. G. O'Connor. 2000. Resource heterogeneity and ungulate population dynamics. *Oikos* 89: 283–294.

Jacobson, A. R., A. Provenzale, A. von Hardenberg, B. Bassano, and M. Festa-Bianchet. 2004. Climate forcing and density dependence in a mountain ungulate population. *Ecology* 85: 1598–1610.

Jewell, P. A., S. Holt, and D. Hart (eds). 1981. *Problems in the Management of Locally Abundant Wild Animals*. Academic Press, New York.

Kaji, K., H. Okada, M. Yamanaka, H. Matsuda, and T. Yabe. 2004. Irruption of a colonizing Sika deer population. *Journal of Wildlife Management* 68: 889–899.

Keep, M. E. 1973. Factors contributing to a population crash of nyala in Ndumu Game Reserve. *Lammergeyer (Natal)* 19: 16–23.

Klein, D. R. 1968. The introduction, increase and crash of reindeer on St. Matthew Island. *Journal of Wildlife Management* 32: 350–367.

Leopold, A. 1943. Deer irruptions. *Wisconsin Conservation Bulletin* 8: 2–11.

McCullough, D. R. 1997. Irruptive behavior in ungulates. In *The Science of Overabundance: Deer Ecology and Population Management*, eds. W. J. McShea, H. B. Underwood, and J. H. Rappole, pp. 69–98. Smithsonian Institution, Washington, DC.

McLaren, B. E. and R. O. Peterson. 1994. Wolves, moose, and tree rings on Isle Royale. *Science* 266: 1555–1558.

McNaughton, S. J. 1984. Grazing lawns, animals in herds, plant form, and coevolution. *American Naturalist* 124: 863–886.

McNaughton, S. J., F. F. Banyikwa, and M. M. McNaughton. 1997. Promotion of the cycling of diet-enhancing nutrients by African grazers. *Science* 278: 1798–1800.

McShea, W. J. 2005. Forest ecosystems without carnivores: when ungulates rule the world. In *Large Carnivores and the Conservation of Biodiversity*, eds. J. C. Ray, K. Redford, R. S. Stenbeck, and J. Berger, pp. 138–153. Island Press, Washington, DC.

McShea, W. J., H. B. Underwood, and J. H. Rappole (eds). 1997. *The Science of Overabundance: Deer Ecology and Population Management*. Smithsonian Books, Washington, DC.

Mduma, S. A. R., A. R. E. Sinclair, and R. Hilborn. 1999. Food regulates the Serengeti wildebeest: a 40-year record. *Journal of Animal Ecology* 68: 1101–1122.

Messier, R. and M. Crete. 1985. Moose-wolf dynamics and the natural regulation of moose populations. *Oecologia* 65: 503–512.

Milchunas, D. G. and W. K. Lauenroth. 1993. Quantitative effects of grazing on vegetation and soils over a global range of environments. *Ecological Monographs* 63: 327–366.

Miller, F. L., S. J. Barry, and W. A. Calvert. 2005. St. Matthew Island reindeer crash revisited: their demise was not nigh – but then, why did they die? *Rangifer*, Special Issue 16: 185–197.

Mott, J. J. 1987. Patch grazing and degradation in native pastures of the tropical savannas in northern Australia. In *Grazing-lands Research at the Plant-animal Interface*, eds. F. P. Horn, J. Hodgson, J. J. Mott, and R. W. Brougham. Winrock International, Virginia.

Mwangi, E. M. 1998. Large herbivore dynamics in the face of insularization: the case of Lake Nakuru National Park, Kenya. *African Journal of Ecology* 36: 276–279.

Mysterud, A. 2006. The concept of overgrazing and its role in management of large herbivores. *Wildlife Biology* 12: 129–141.

NRC (National Research Council). 2002. *Ecological Dynamics on Yellowstone's Northern Range*. National Academy Press, Washington, DC.

Ogutu, J. O. and N. Owen-Smith. 2005. Oscillations in large herbivore populations: are they related to predation or rainfall? *African Journal of Ecology* 43: 332–339.

Owen-Smith, N. 1988. *Megaherbivores. The Influence of Very Large Body Size on Ecology*. Cambridge University Press, Cambridge.

Owen-Smith, N. 2000. Modeling the population dynamics of a subtropical ungulate in a variable environment: rain, cold and predators. *Natural Resource Modeling* 13: 57–87.

Owen-Smith, N. 2002a. *Adaptive Herbivore Ecology*. Cambridge University Press, Cambridge, UK.

Owen-Smith, N. 2002b. A metaphysiological modelling approach to stability in herbivore-vegetation systems. *Ecological Modelling* 149: 153–178.

Pastor, J. and Y. Cohen. 1997. Herbivores, the functional diversity of plants species, and the cycling of nutrients in ecosystems. *Theoretical Population Biology* 51: 165–179.

Pastor, J., Y. Cohen, and N. T. Hobbs. 2006. The roles of large herbivores in ecosystem nutrient cycles. In *Large Herbivore Ecology, Ecosystem Dynamics and Conservation*, eds. K. Danell, R. Bergstrom, P. Duncan, and J. Pastor, pp. 289–325. Cambridge University Press, New York.

Peterson, R. O. 1999. Wolf-moose interaction on Isle Royale: the end of natural regulation? *Ecological Applications* 9: 10–16.

Peterson, R. O., J. A. Vucetich, R. E. Page, and A. Chouinard. 2003. Temporal and spatial aspects of predator-prey dynamics. *Alces* 39: 215–232.

Post, E., N. C. Stenseth, R. O. Peterson, J. A. Vucetich, and A. M. Ellis. 2002. Phase dependence and population cycles in a large-mammal predator-prey system. *Ecology* 83: 2997–3002.

Putkonen, J. and G. Roe. 2003. Rain-on-snow events impact soil temperatures and affect ungulate survival. *Geophysical Research Letters* 30(4) article 1188: 1–4.

Reed, D. H., J. J. O'Grady, J. D. Ballou, and R. Frankham. 2003. The frequency and severity of catastrophic die-offs in vertebrates. *Animal Conservation* 6: 109–114.

Riney, T. 1964. The impact of introductions of large herbivores on the tropical environment. *International Union for the Conservation of Nature, Publication New Series* 4: 261–273.

Ritchie, M. E., D. Tilman, and J. M. H. Knops. 1998. Herbivore effects on plant and nitrogen dynamics in oak savanna. *Ecology* 79: 165–177.

Scheffer, V. B. 1951. The rise and fall of a reindeer herd. *Science Monthly* 73: 356–362.

Schmitz, O. J. and A. R. E. Sinclair. 1997. Rethinking the role of deer in forest ecosystem dynamics. In *The Science of Overabundance: Deer Ecology and Population Management*, eds. W. J. McShea, H. B. Underwood, and J. H. Rappole, pp. 201–223. Smithsonian Books, Washington, DC.

Simard, M. A., S. D. Côté, R. B. Weladji, and J. Huot. 2008. Feedback effects of chronic browsing on life-history traits of a large herbivore. *Journal of Animal Ecology* 77: 678–686.

Sinclair, A. R. E. 1979. The eruption of the ruminants. In *Serengeti: Dynamics of an Ecosystem*, eds. A. R. E. Sinclair and M. Norton-Griffiths, pp. 82–103. University of Chicago, Chicago, Ill.

Singer, F. J. and J. E. Norland. 1994. Niche relationships within a guild of ungulate species in Yellowstone National Park, Wyoming, following release from artificial control. *Canadian Journal of Zoology* 72: 1383–1394.

Toigo, C., J.-M. Gaillard, M. Festa-Bianchet, E. Largo, J. Michallet, and D. Maillard. 2007. Sex- and age-specific survival of the highly dimorphic Alpine ibex: evidence for a conservative life-history tactic. *Journal of Animal Ecology* 76: 679–686.

Tremblay, J.-P, I. Thibault, J. Huot, and S. D. Cote. 2005. Long-term decline in white-tailed deer browse supply: can lichens and litterfall act as alternative food sources that preclude density-dependent feedbacks?. *Canadian Journal of Zoology* 83: 1087–1096.

Turchin, P. 2003. *Complex Population Dynamics*. Princeton University Press, Princeton, NJ.

Turchin, P. and G. O. Batzli. 2001. Availability of food and the population dynamics of arvicoline rodents. *Ecology* 82: 1521–1534.

Valkenburg, P., R. A. Sellers, D. C. Squibb, J. D. Woolongton, A. R. Aderman, and B. W. Dale. 2003. Population dynamics of caribou herds in southwestern Alaska. *Rangifer* Special Issue No. 14, 131–142.

Vucetich, J. A. and R. O. Peterson. 2004a. The influence of top-down, bottom-up and abiotic factors on the moose (*Alces alces*) population of Isle Royale. *Proceedings of the Royal Society of London Series B: Biological Sciences* 271: 183–189.

Vucetich, J. A. and R. O. Peterson. 2004b. Long-term population and predation dynamics of wolves on Isle Royale. In *Biology and Conservation of Wild Canids*, ed. D. Macdonald and C. Sillero-Zubiri, pp. 281–292. Oxford University Press, Oxford.

Wagner, F. 2006. *Yellowstone's Destabilized Ecosystem*. Oxford University Press, New York.

Walker, B. H., R. H. Emslie, R. N. Owen-Smith, and R. J. Scholes. 1987. To cull or not to cull: lessons from a southern African drought. *Journal of Applied Ecology* 24: 381–401.

Ward, D. 2006. Long-term effects of herbivory on plant diversity and functional types in arid ecosystems. In *Large Herbivore Ecology, Ecosystem Dynamics and Conservation*, eds. K. Danell, R. Bergstrom, P. Duncan, and J. Pastor, pp. 142–169. Cambridge University Press, New York.

White, P. J., J. E. Bruggeman, and R. A. Garrott. 2007. Irruptive population dynamics in Yellowstone pronghorn. *Ecological Applications* 17: 1598–1606.

Wilmers, C. C., E. Post, and A. Hastings. 2007. The anatomy of predator-prey dynamics in a changing climate. *Journal of Animal Ecology* 76: 1037–1044.

Woodgerd, W. 1964. Population dynamics of bighorn sheep on Wildhorse Island. *Journal of Wildlife Management* 28: 381–391.

Yellowstone National Park. 1997. *Yellowstone's Northern Range: Complexity and Change in a Wildland Ecosystem.* U.S. National Park Service, Mammoth Hot Springs, Wyoming.

6

How does landscape heterogeneity shape dynamics of large herbivore populations?

N. Thompson Hobbs[1] and Iain J. Gordon[2]

[1]*Natural Resource Ecology Laboratory and Graduate Degree Program in Ecology, Colorado State University, Fort Collins, Colorado, United States of America*

[2]*CSIRO Sustainable Ecosystems, Davies Laboratory, Aitkenvale, Queensland, Australia*

A central challenge in ecology seeks to understand the dynamics of populations. It is clear that the abundance of organisms varies over time and space in response to two broad sets of processes: those that operate within populations and those that are external to them. Internal controls on dynamics include feedbacks from the current state of the population to its future state; for example, demography and population density shape rates of population growth. External controls on dynamics, particularly weather and its effect on resources, include conditions that are not influenced by characteristics of the population. Efforts to describe the role of these controls on population dynamics have dominated ecological inquiry for decades (Kingsland 1985).

Until recently, virtually all simple models of herbivore population dynamics have ignored trophic interactions entirely (i.e., Chapter 2) or have borrowed from abstractions of predator–prey interactions, representing feedbacks from plants to herbivores based on a functional response of

Dynamics of Large Herbivore Populations in Changing Environments, 1st edition. Edited by Norman Owen-Smith.
© 2010 Blackwell Publishing

herbivores to the quantity of plants available to them (Noy Meir 1975, Caughley 1976, Caughley and Lawton 1981, Caughley 1982, Schmitz 1993, Blatt et al. 2001). Variance in plant resources, particularly variation in plant quality, has been usually ignored in these models despite the clear importance of spatial heterogeneity in shaping dynamics of predator–prey and host–parasitoid systems (see reviews of Hassell 1980, Hassell and Pacala 1990, Holt and Hassell, 1993). Because variation in quality and quantity are characteristic of the foods consumed by herbivores, it follows that mathematical theory on plant–herbivore interactions may fail to represent critical controls on herbivore population dynamics (but see Edelstein-Keshet 1986, Owen-Smith 2002a, b).

The preponderance of empirical studies of population dynamics has also overlooked spatial variation in plant resources. Investigations of population responses to resources have most often focused on how the total or average quantity of resources mediates population dynamics by constraining equilibrium population size or influencing population growth rate (Merrill and Boyce 1991, Fabricius 1994, Kiker 1998, Mduma et al. 1999, Weisberg et al. 2002). Moreover, although many time-series analyses have studied effects of weather and population covariates on population behavior (Putman et al. 1996, Forchhammer et al. 1998, Mysterud et al. 2000, Jacobson et al. 2004, Mysterud and Ostbye 2006, Colchero et al. 2009), most of these studies focused exclusively on explaining variation in time, averaging over the heterogeneity that exists in space.

Population dynamics of large herbivores unfold in a spatial context. In this chapter, we explore recent work revealing how populations of ungulate herbivores respond to spatial heterogeneity in resources, particularly heterogeneity in plant quality. We depart from earlier studies in an important way. Historically, investigations of the effect of spatial heterogeneity on herbivores have focused on the ways whereby individual foraging animals discriminate among spatially variable resources arrayed across a range of spatial scales (see reviews of Senft et al. 1987, Laca and Demment 1991, Bailey et al. 1996, Hobbs 1999). Although individual choices clearly aggregate to influence populations, population-level responses of large herbivores to spatial heterogeneity in resources have not been described until recently (Illius and O'Connor 2000, Wang et al. 2006, Hebblewhite et al. 2008, Post and Forchhammer 2008, Post et al. 2008, Wang et al. 2009). Here, we consider the ways that differences in the levels of heterogeneity within landscapes influence population dynamics and we propose mechanisms explaining these influences.

Our chapter will be organized as follows. We begin by defining some terms. Next, we summarize theoretical and empirical evidence showing that population trajectories of ungulates are shaped by spatial heterogeneity in resources. We then outline mechanisms explaining why heterogeneous environments influence population dynamics, and we review empirical evidence for the operation of these mechanisms. We close by considering how global change may affect ungulate populations by limiting the access of individuals to heterogeneity in resources.

6.1   What is spatial heterogeneity?

Ecologists use the term "heterogeneity" with a bewildering diversity of meanings (Kolasa and Rollo 1991, Hobbs 2003), so we begin by offering an operational definition. Adler et al. (2001) pointed out that when spatial heterogeneity is measured using aspatial statistics, it is synonymous with spatial variability, but when it is measured with spatially explicit statistics, it implies spatial dependence or spatial pattern. Thus, heterogeneity, taken to mean variability over space, can be evaluated with standard deviations or coefficients of variation, while heterogeneity, taken to mean spatial dependence, must be measured by spatially explicit statistics like spatial autocorrelation or geostatistics (e.g., Pastor et al. 1998, Pastor et al. 1999). Although there is a growing literature on responses of individual foragers to spatial pattern (Hobbs 1999, Hobbs et al. 2003, Searle et al. 2005, Searle et al. 2006), little is known about the ways whereby spatial pattern influences population dynamics. Thus, for the purposes here, heterogeneity can be equated with variability over space, facilitating comparisons with variability measured over time. We will use the standard deviation or coefficient of variation of resources taken from samples arrayed over space (e.g., pixels in a landscape) as an operational measure of heterogeneity. There will be cases where autocorrelation in time or space will be important and we will be careful to highlight these instances. However, throughout this chapter, our use of heterogeneity will imply the following: a system that is spatially heterogeneous will show high coefficients of variation in a variable of interest (for example, plant biomass, plant nitrogen content) sampled from many locations, while a spatially homogeneous system will show low coefficients of variation. Here, we are referring to heterogeneity at the landscape scale comprising variation between habitats in resource attributes. For our purposes, we will use the term landscape to mean the area used by a population.

6.2   How does spatial heterogeneity influence ungulate population dynamics?

Consider two landscapes with similar levels of net primary production. One of these landscapes is spatially heterogeneous – it contains a varied array of vegetation types, topographic positions, and elevations. The other lacks this heterogeneity. How will herbivore populations respond to these different spatial contexts? Answering this question empirically is challenging because doing so requires observations of population trajectories on each landscape. Simply observing each population at a few points in time in each location will not suffice. Understanding how a population responds to "variation in spatial variation" is even more difficult because we need to observe population behavior over a range of landscape types varying in their levels of spatial heterogeneity. As a result of these difficulties, empirical studies of responses of populations to landscape heterogeneity have been infrequent. However, despite these difficulties, emerging evidence suggests that spatial heterogeneity in landscapes enhances performance of populations of large herbivores.

6.2.1   Results from analytical and simulation models

The formidable challenges of assembling data on effects of heterogeneity on population performance has not deterred a rich set of modeling efforts focused on understanding these effects. With few exceptions, these models have relied on relatively detailed, species-specific simulations; analytically based findings are less common. Predictions emerging from these models include the following:

1 Heterogeneity in resources that buffer against shortages during the dormant season can increase the long-term, average abundance of herbivore populations (Illius and O'Connor 2000, Owen-Smith 2002b).

2 These buffering resources can enhance the long-term stability of populations (Illius and O'Connor 2000, Owen-Smith 2002a, b, 2004). Whether such resources are stabilizing depends on their quality and on critical thresholds for herbivore starvation (Owen-Smith 2002a, b, 2004). In particular, the buffering resources must be of intermediate quality. In this case, animals do not starve during periods of resource scarcity but simply lose condition, thereby limiting reproduction and reducing the tendency of the population to overshoot its equilibrium (Illius 2006).

3 Movement of large herbivores among patches of resources within landscapes, patches that vary over time and space, can enhance population growth rates (Fryxell et al. 2005), increase supportable densities of animals (Boone and Hobbs 2004, Underwood 2004, Boone et al. 2005, Boone 2007), and promote persistence of populations that in the absence of spatial heterogeneity would go extinct (Fryxell et al. 2005). These effects depend on low spatial autocorrelation in patch quality and herbivore mobility (Fryxell et al. 2005). If spatial autocorrelation is high or mobility is restricted, enhancing effects of resource heterogeneity on population performance are lost.

4 Effects of heterogeneity depend on overall levels of resource abundance. Resource heterogeneity exerts its greatest effect at intermediate levels of resource abundance (Boone 2007).

5 Heterogeneity in resource types can foster coexistence among animals differing in body mass and feeding style, and, so doing, enhance the diversity of herbivore communities and increase secondary production (Owen-Smith 2002b).

6 Spatial variation in the resources can enhance the abundance of consumers (predators or herbivores), whenever feedbacks from consumers to resource production are weak, and relationships between consumer abundance and the quantity of resources are nonlinear (Doncaster 2001). The enhancing effect of heterogeneity on consumer abundance is opposed by density dependence in the exploitation of resources created by interference competition. It may be dampened or reversed when efficiency of exploitation of resources by consumers is high, when there are strong feedbacks to resource production, or when the relationship between consumer abundance and resource quantity is linear (Doncaster 2001).

7 Feedbacks from herbivores to the distribution of forage quality can create density dependence apart from any effects of herbivores on forage quantity (Edelstein-Keshet 1986).

Clearly, none of these predictions would emerge from simple models of herbivore population dynamics (Chapter 2) that fail to represent heterogeneity in plant quantity and quality. However, although these predictions are new and offer useful motivation for empirical tests, they have not yet been formed into a comprehensive, integrated theory on responses of herbivore populations to heterogeneity, but rather provide a series of vignettes where a plausible set of premises leads by induction to a set of outcomes. These outcomes clearly add to our understanding of herbivore population dynamics, but they have not yet been brought together in a comprehensive theory.

Some of these models, notably the metaphysiological models of Owen-Smith (2002b), are justified as alternatives to far simpler models, like the logistic and Ricker equations and others that model the approach of populations to equilibrium (Chapter 2). This appears to be a false comparison, because the purposes of the types of models are fundamentally different – a model cannot simultaneously be simple and, therefore, general, and at the same time, be more detailed and, therefore, specific (Levins 1966).

Although there is no question that the detailed models described above provide greater fidelity to specific conditions, they do so at a price. For detailed models, it remains uncertain whether the outcomes are constrained by the specific set of parameter values chosen for the simulations, and the extent to which the "generalities" that emerge are limited to the specific systems motivating the choice of those values.

6.2.2 Empirical results

Some of the predictions of models have been corroborated by empirical studies. Wang et al. (2006) examined time series of estimates of abundance of elk and bison in western North America and evaluated the strength of density dependence operating in each population. Strength of density dependence was estimated from the slope of annual per capita growth rate as a function of annual estimates of population size in a discrete time, Gompertz model of population growth. Increasingly negative slopes indicated stronger feedbacks from population density to population growth rate. The authors developed indices of temporal heterogeneity using the coefficient of variation in winter temperatures and developed indices of spatial heterogeneity using the coefficient of variation in the Normalized Difference Vegetation Index (NDVI). Temporal heterogeneity in winter weather amplified feedbacks from population density to population growth rate, strengthening density dependence, but spatial heterogeneity in NDVI weakened these feedbacks. Because the x-intercept of the slope of per capita population growth rate on population size estimates the potential population density at equilibrium, the result of Wang et al. (2006) implied that spatially heterogeneous landscapes could support greater densities of animals at equilibrium than landscapes that were less heterogeneous. In the same way, temporal heterogeneity reduced the equilibrium population size. Thus, a pivotal result emerging from this work was that heterogeneity in time and space acted in ways that were diametrically opposed (Fig. 6.1).

Similarly, Wang et al. (2009) estimated the strength of density dependence in populations of ungulates in Europe and North America occupying landscapes with different levels of spatial variance in elevation. They observed a negative correlation between spatial heterogeneity in altitude and the strength of density dependence, reinforcing their earlier findings. Average elevation was unrelated to strength of density dependence, leading to the conclusion that variation in resources along altitudinal gradients, rather than main effects of altitude, were responsible for effects on population dynamics. Wang et al. (2009) surmised that variation in altitude offered a surrogate for variation in plant phenology, which,

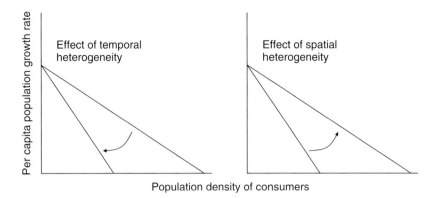

Figure 6.1 Opposing effects of temporal and spatial heterogeneity on productivity of populations of ungulates observed by Wang et al. (2006). *Arrows* represent influence of increasing heterogeneity. Changes in the slopes of the lines represent effects on the strength of density-dependent feedback to population growth – steeper slopes mean stronger feedback. Because the x-intercept of the lines is landscape-carrying capacity, these hypothesized relationships predict that temporal heterogeneity reduces the number of consumers that can be supported, while spatial heterogeneity increases it.

in turn, enhanced nutritional status of individuals as described in the subsequent section.

Post and Stenseth (1999) studied reproductive performance of female red deer in relation to spatial heterogeneity in the number of calves born during years with spatial heterogeneity in date of flowering of vegetation. Females born in years when spatial variation in flowering date was high were about 25% more likely to bear calves as 2-year olds then were females born when heterogeneity in flowering was low (Post and Stenseth 1999). Langvatn et al. (1996) found that age of first reproduction in red deer in Norway and on the Isle of Rum, Scotland was influenced by spring temperatures; warm temperatures were associated with delayed reproduction, cool springs with accelerated reproduction. The authors attributed this effect to slowing of plant phenology by cold weather and consequent increased spatial variation in plant age. This variation, in turn, was believed to allow greater opportunity for females to feed on highly nutritious plant tissue by expanding the window of time when animals could consume plant tissue of peak quality.

Offspring production by caribou increased in direct proportion to spatial variation in plant phenology at 1–100 m scales in West Greenland (Post et al. 2008) and was inversely related to the extent of asynchrony between the timing of peak calf production and the timing of peak emergence in plants (i.e. "trophic mismatch," Post and Forchhammer 2008,

Post et al. 2008). Both relationships offer evidence that access to spatial heterogeneity in plants enhances reproductive performance.

6.3 Mechanisms explaining the influence of spatial heterogeneity on population dynamics

In this section, we describe mechanisms that may be responsible for the influence of spatial heterogeneity on population dynamics and offer evidence for the operation of these mechanisms. Population behavior is the aggregate outcome of fates of individuals. Because mechanistic explanations for higher levels in ecological hierarchies require reference to lower level behavior (O'Neill et al. 1986), explaining population patterns mechanistically requires explanations of processes affecting individuals.

We offer two explanations for why populations respond to heterogeneity as described above. Results showing reduced strength of density dependence and enhanced vital rates can be explained by what we will call mechanism 1: spatial heterogeneity prolongs the period of time during which animals can exploit vegetation at peak nutritional quality. Effects on density dependence and population stability can be explained by mechanism 2: spatial heterogeneity can create buffering resources that reduce mortality when temporal heterogeneity in resources creates episodic shortages of resources.

6.3.1 Spatially heterogeneous landscapes enhance opportunities to exploit high-quality resources

Nutritional benefits accrue to ungulates that exploit heterogeneity in plant phenology (Albon and Langvatn 1992, Frank and McNaughton 1992, Wilmshurst et al. 1999, Mysterud et al. 2001a, Hebblewhite et al. 2008). In this section, we develop a general theoretical framework showing why mobility by ruminant herbivores in heterogeneous environments offers nutritional benefits to individuals – benefits that can translate into benefits for population performance. This framework begins with the idea that increases in biomass of individual plants and plant communities that occur as plants grow are correlated with reductions in plant nutritional quality (Deinum et al. 1981, Demment and Van Soest 1985, McNaughton 1990, Hume 1991, Coblentz et al. 1998, Van der Wal et al. 2000, Gustavsson and Martinsson 2004, Hebblewhite et al. 2008). This means that there is a positive relationship between phenological progression and plant biomass and a negative relationship between advancing phenology and plant nutritional value. These relationships create important trade-offs for foraging herbivores – trade-offs that are mediated through foraging and

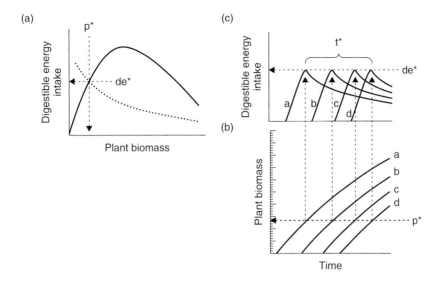

Figure 6.2 (a) Digestible energy (de) intake is limited by foraging (*solid line*) or digestion (*dotted line*), whichever is less. There is a biomass of plants (p*) that provides the maximum daily de intake. (b) Consider four habitats (*a–d*) with asynchronous phenology caused by delays in the initiation of plant growth with habitat *a* being earliest and *d* latest. Heterogeneity in phenology means that p* occurs at different times in different locations. (c) As a result, the limiting portions of the foraging and digestion constraint curves show peaks (de*) at different times. These differences expand the time interval (t*) when herbivores can obtain maximum digestible energy. When plant growth is synchronous, the growth curves in panel (b) move close together, compressing t*.

digestive processes constraining daily energy intake (Fryxell 1991, Shipley et al. 1999).

Realized intake of plants by herbivores is the lesser of foraging and digestive constraints (Fig. 6.2a). Consider a landscape that contains a set of habitats with asynchronous phenology (Fig. 6.2b); spatial heterogeneity is determined by the extent of correlation in phenology among habitats – highly correlated habitats are synchronous in time and homogeneous in space; habitats with low correlation are asynchronous in time and heterogeneous in space. Movement among asynchronous habitats allows animals to maintain maximum energy intake by selecting habitats with the optimum biomass for achieving energy intake. Thus, heterogeneity in phenology prolongs the interval of time during which animals can obtain the maximum potential energy gain (Fig. 6.2c).

Reductions in landscape heterogeneity or reductions in herbivore mobility among habitats compress the interval of time over which animals can obtain maximum potential energy gain. If habitats that are asynchronous in their phenological response are arrayed along a spatial gradient, then animals should migrate along the phenological gradient, maintaining maximum nutrient intake rate (Fig. 6.2b, c). When plants are senescent, biomass and quality become effectively uncoupled – because plant tissue is of similar quality in all habitats. In this case, animals should be expected to migrate up the biomass gradient, because high biomass forage supplies can withstand higher levels of grazing before they become depleted. These kinds of movements also create buffering resources by assuring that some biomass remains unconsumed, a phenomenon described in the next section.

6.3.2 Spatially heterogeneous landscapes contain buffering resources

Virtually all populations of large herbivores experience periodic shortages in resources when plants are dormant, during the dry season in tropical environments and during the winter in temperate ones. Moreover, the total amount of edible biomass that is produced during any single growing season can vary enormously among years. Annual and seasonal variation in the availability of resources has the potential to cause dramatic fluctuations in animal numbers in the absence of resources that remain stable among seasons and years (e.g., Caughley et al. 1985, Ellis and Swift 1988, Fletcher et al. 1990, Owen-Smith 2004).

Heterogeneous landscapes contain a diversity of soil types, hydrologic features, and topographic positions, which, in turn, create spatial variation in standing crop biomass and vegetative composition. This diversity often includes landscape positions where plant biomass remains available for consumption by herbivores even when shortages prevail in other areas of the landscape. For example, tall grasslands in the northern Serengeti of East Africa provide biomass during the dry season when southern short grasslands are largely depleted (McNaughton 1983, Boone et al. 2006). In north-temperate ecosystems, habitats at low elevation remain largely snow free when high elevation habitats offer virtually no food because of deep snow (Wallmo et al. 1977, Frank 1998). In addition, heterogeneous landscapes contain a diversity of plant species and growth forms that respond to temporal variation in weather in different ways. Deeply rooted shrubs can persist through droughts, providing leaves when the herb layer fails to grow as a result of moisture stress (Scholes and Archer 1997). Plants with erect growth forms protrude above accumulated snow when prostrate ones are completely covered (Johnson et al. 2001, Nordengren et al. 2003). Such resources can permit herbivores to survive resource shortages, buffering populations from temporal heterogeneity in plant production and availability (Hobbs 1989, Illius and O'Connor 2000).

A key part of this buffering is spatial variation in plant quality that allows animals to forgo consumption of low-quality resources when overall resource abundance is high (Owen-Smith 2002a, b, 2004, Illius 2006). If herbivores avoid consuming low-quality plants when high-quality ones are abundant, then low-quality biomass can accumulate to offset episodes of resource shortage. However, these accumulating resources must be of sufficient quality to prevent starvation during the dormant season. The phenomena of buffering may be common along productivity gradients. For the reasons described above, there is an inverse relationship between nutritional quality and plant biomass. Thus, landscapes with gradients in moisture often create opposing gradients in plant nutritional quality and biomass. The high productivity end of these gradients offers buffering resources that are avoided when high-quality resources are abundant and exploited when high-quality resources are rare.

6.4   Influences from high-quality resources

Enhancing access to nutritious forage forms an important selective force in the evolution of migration in large herbivores (Fryxell et al. 1988, Mysterud et al. 2001a). Spatiotemporal variation in plant phenology entrains movements of large herbivores in ecosystems throughout the world, movements that are qualitatively consistent with mechanism 1. There is a common trend over time and space – animals migrate down phenological gradients during the growing season, tracking biomass and quality optima, and then move to areas of low quality, high biomass during periods of plant dormancy (McNaughton 1979, 1990).

In the Serengeti region of East Africa, zebra and wildebeest migrate over a distance spanning 200 km, following a rainfall gradient that shapes plant phenology and quality (McNaughton 1979, 1990, Boone et al. 2006). Similar patterns of movement along moisture/phenology gradients have been observed for wild and domestic ruminants in other African ecosystems (Breman and Wit 1983, Sinclair and Fryxell 1985, Fryxell and Sinclair 1988). Models of movement of wildebeest in the Serengeti (Boone et al. 2006, Holdo et al. 2009) were able to mimic observed migration patterns of wildebeest on the basis of spatial variation in rainfall, gradients in plant nutritional quality, and availability of young, growing vegetation. Simple models predicting that animals will maximize their energy intake by tracking phenologically young patches of grass in the Serengeti have successfully explained the spatial distribution of wildebeest

(Wilmshurst et al. 1999) and Thompson's gazelles (Fryxell et al. 2004) across a heterogeneous landscape.

In temperate systems, snowmelt and plant emergence appear to be the primary cues for migration from low to high elevation. Initiation of plant growth explained initiation of annual variation in migration of red deer in Norway (Pettorelli et al. 2005). Elk populations in Yellowstone National Park migrate along an elevation gradient, apparently tracking variation in plant nitrogen content and the concentration of green biomass in grass patches (Frank and McNaughton 1992). In so doing, these animals gain access to immature plants for up to 6 months of the year (Frank and McNaughton 1992). In the absence of migration along an elevation gradient, this access would likely be compressed to an interval of weeks (Baker and Hobbs 1982). Elk in the Rocky Mountains of Canada migrate to areas of high topographic diversity, allowing them to exploit forage patches of intermediate biomass and high forage quality (Hebblewhite et al. 2008). As a result, migratory elk obtained diets that were 6.5% points higher in dry matter digestibility than diets chosen by nonmigratory elk (Hebblewhite et al. 2008). Caribou in West Greenland track phenology of emerging plants while foraging. The timing of their reproduction appears to be synchronized with plant phenology to allow maximum opportunity to forage on newly emerging and spatially heterogeneous plant biomass when calves are produced (Post and Forchhammer 2008).

Phenology has also been implicated as a primary determinant of movements by large herbivores in the steppes of Mongolia. Mueller et al. (2008) used NDVI to explain spatial distribution of Mongolian gazelles and found that intermediate values of NDVI were most predictive of their movements. This suggests that animals were influenced by the trade-off between plant biomass and quality, selecting landscape positions where plants were of intermediate maturity and biomass. Similar to Thompson's gazelles in East Africa (Fryxell et al. 2004), Mongolian gazelles in Asia exploit spatial heterogeneity in plants arrayed over enormous areas of landscape (Mueller et al. 2008).

Accumulating evidence shows that these types of movements in heterogeneous environments enhance animal condition. For example, body mass of red deer was positively correlated with increasing positive values of the North Atlantic Oscillation (NAO) (Mysterud et al. 2002, Pettorelli et al. 2003), a correlation explained by increased spatial heterogeneity in plant phenology on high elevation summer ranges, heterogeneity that was amplified by increased snowfall in a topographically complex landscape. Spatiotemporal variability in phenology, in turn, was believed to prolong access of animals to high-quality forage relative to winters when NAO was low and snowfall was reduced at high elevations (Mysterud et al. 2001b, Pettorelli et al. 2003). This explanation was reinforced by the observation of elevated body mass of red deer that migrated seasonally along an elevation gradient (corresponding to a gradient in delayed plant phenology and increased plant protein content) relative to body mass of red deer that remained year-round in low-elevation habitats (Albon and Langvatn 1992). Similarly, body mass of moose was positively related to weather patterns that retarded phenological development of forage and prolonged access of moose to young forage (Herfindal et al. 2006).

The most direct evidence to date of beneficial effects of spatiotemporal heterogeneity in phenology on animal condition was offered by Mysterud et al. (2001a), who studied the influence of altitudinal migration on body mass of red deer in Norway. Migration along an elevation gradient can offer nutritional benefits to herbivores in two ways — by providing access to forage that grows more slowly due to colder conditions on north slopes and at high elevation, or by providing access to a topographically complex landscape with heterogeneous phenological trajectories. Mysterud et al. (2001a) sought to tease apart the effect of slow growth of plants that occurs at high altitude and on north slopes from the effect of heterogeneity in altitude and aspect. They observed a positive correlation between body mass and heterogeneity in elevation and aspect but failed to find an effect of high elevation or north facing slopes per se. This showed that the benefits of migration result from landscape heterogeneity that creates a diversity of phenologies rather than from conditions that simply slow phenological progression. Moreover, the effect of altitude alone, separate from effects of spatial variability, has been shown to be *inversely* correlated with body mass of nonmigratory moose (Hjeljord and Histol 1999, Ericsson et al. 2002), reinforcing the conclusion of Mysterud et al. (2001a) that heterogeneity, rather than altitude of mountainous terrain, is responsible for nutritional benefits of migration.

Although the work reviewed above suggests that spatial heterogeneity in phenology enhances animal condition and population performance, there are potential trade-offs relative to homogeneously distributed resources. Spatial heterogeneity in resources implies that there are "hotspots" — areas of the landscape where asynchronous resources are much more valuable to consumers, at any one time, than the average of the landscape. Indeed, successful models of animal movement in heterogeneous landscapes depend on this difference between the average and specific locations to motivate animal movement (e.g., Fryxell et al. 2004). Attraction to these spatially concentrated resources causes aggregation (Fryxell 1991). Aggregation of herbivores, in turn, will presumably increase grazing pressure that may lead to the development of grazing lawns that promote further aggregation (McNaughton 1989, Jefferies et al. 1994), thereby amplifying intraspecific competition for resources (Murray and Illius 2000).

The operation of this trade-off may depend on temporal heterogeneity in weather; for example, increasing effects of NAO on snow cover in Greenland appear to dictate the spatial dispersion of muskoxen (Forchhammer et al. 2005). High, positive NAO winters were associated with reduced snow accumulation during winter which increased the average annual plant biomass and the degree of spatial synchrony in plant growth in the following summer. This increased spatial synchrony in plant growth reduced aggregation of muskoxen relative to years when biomass and spatial synchrony were low. Spatial dispersion of individuals across landscapes during high NAO years may allow muskoxen to reduce competition for food while maintaining access to large quantities of high-quality forage (Forchhammer et al. 2005). Thus, although benefits of spatial heterogeneity induced by asynchronous phenology to the individual forager are clear, these benefits become less certain when animals forage in groups. It follows that understanding effects of spatial heterogeneity at levels of organization above the individual will involve aggregated consumer demand and its effects on resource supply and renewal. This empirical result is the first to represent the theoretical prediction of Doncaster (2001), that benefits of heterogeneity are opposed by intraspecific competition for resources.

6.5   Influences from buffer resources

In both tropical and temperate ecosystems, herbivores have been shown to move from consuming high-quality resources during the growing season into habitats that contain more abundant resources during the dry or dormant season. These habitats tend not to be used in the growing season because their vegetation is of lower quality than that available in other habitats. This movement may be localized or small scale (Bell 1971, Gordon 1989) or may involve migration such as in the case of wildebeest moving from the short grass plains in the Serengeti used in the wet season to the high biomass grasslands of the Masai Mara in the dry season (Fryxell et al. 2004).

At the vegetation community scale, Gordon (1989) found that cattle and ponies on the Isle of Rum selected for vegetation communities that contained high-quality vegetation during the growing season, but moved off these communities onto vegetation communities containing higher biomasses of lower quality vegetation during the winter. The smaller bodied goats and red deer within this guild foraged on higher quality vegetation throughout the year. This indicates that the extent to which species use alternative resources in the growing season and the dormant season depends on the species body size and digestive system. Large-bodied species and those with a hindgut fermentation system can deal with poorer quality plant material and have high total intake requirements (Illius and Gordon 1991, 1992) and, therefore, move off higher quality resources earlier than small-bodied species (also driven by competition; Illius and Gordon 1987). Similarly, Bell (1971) showed that the guild of herbivores in the Serengeti (zebra, wildebeest, and Thompson's gazelle) moved up and down a catena gradient, using the upper part of the catena during the wet season when highly nutritious grasses and forbs were available and moved down to the lower, wetter parts of the catena to feed on more abundant, lower quality grasses during the dry season.

In western North America and Northern Europe, native ungulates used high-elevation habitats during the growing season and low-elevation habitats during winter (Garrott et al. 1987, Loft et al. 1989, Albon and Langvatn 1992, Mysterud et al. 2001a). The ability to migrate among these habitats is critical to population performance because snow accumulation during winter can make forage on productive summer ranges inaccessible, reducing their ability to support animals to virtually zero (Wallmo et al. 1977). Systems like these illustrate the operation of stabilizing heterogeneity (sensu Owen-Smith 2002a, b, 2004). Because production is low and forage quality tends to decline rapidly in low-elevation systems, they are avoided by large herbivores during the growing season. However, during most winters, they offer forage of sufficient amount and quality to largely prevent starvation of adults, and reduce starvation of young animals (Hobbs 1989).

6.6   Global change and access to heterogeneity by large herbivores

We have developed the case that access to spatial heterogeneity in plants influences population dynamics of large herbivores by expanding their options for selective foraging. Two aspects of global change, habitat fragmentation and climate change, may dramatically constrain these options.

Habitat fragmentation, the splitting of intact landscapes into spatially isolated parts, has emerged as a fundamentally important source of environmental change worldwide (Galvin et al. 2008, Hobbs et al. 2008, Wilcove and Wikelski 2008). Historically, the term habitat fragmentation was used to describe two effects that occur simultaneously – habit loss that occurs from conversion of one land cover type to another, and habitat isolation that occurs as movement of organisms among patches of habitat becomes increasingly restricted (Fahrig 2002). In order to separate the often confounded effects of isolation and habitat loss, contemporary landscape ecologists reserve the term *fragmentation* to refer specifically to the isolating effects of changes in landscapes in the absence of a reduction in habitat area (Fahrig 2002). We will use fragmentation in this sense.

A key, unresolved question in landscape ecology asks, "Does the effect of habitat fragmentation add to the effect of habitat loss?" This question can be usefully rephrased as follows: if a landscape is subdivided into a set of isolated parts, with no change in the area of habit within the landscape, what effect should we expect for individual organisms and for populations? In the preceding sections, we showed that access to heterogeneity in resources in landscapes offers nutritional benefits to individual herbivores, benefits that appear to translate into enhanced performance of herbivore populations. These benefits occur because mobility allows herbivores to track peaks in potential maximum energy intake that occur at different times and locations across the landscape. In addition, mobility allows herbivores to exploit resources that are stable in time during episodes of resource shortage. An overlooked effect of fragmentation on organisms is that it compresses the scale of interaction between consumers and resources; that is, consumers that occupy fragmented landscapes have restricted access to the full range of temporal and spatial variation in resources that is available in intact landscapes (Hobbs et al. 2008). The theoretical framework we developed illustrates that fragmentation of habitats can lead to fundamentally important changes in population performance, even when the total area and quality of habitat remains unchanged. If habitat fragmentation reduces this mobility, then animals are compelled to consume forages that diminish in quality with time. Fryxell et al. (2005) and Owen-Smith (2004) showed that reducing the scale at which herbivores can interact with resources that vary over time and space can have profound consequences for population viability, even for an abundant species.

A warming climate and increased climate variability accompany habitat fragmentation as a prevailing source of human-caused change in the global environment. At landscape scales, climate warming appears to amplify spatial heterogeneity in plant growth, and in this way potentially enhances access to high quality forage by large herbivores (Post and Stenseth 1999, Post et al. 2003). However, the increase in heterogeneity over the landscape can be counteracted by reduced heterogeneity in a suite of plant characteristics at finer scales (Post et al. 2008). The reduction in heterogeneity at the spatial scales most relevant to daily foraging as a consequence of warmer temperature conditions was associated with lowered offspring production by female caribou in Greenland (Post et al. 2008). This result suggests that climate warming may reduce heterogeneity in phenology at the scales that are most relevant to daily foraging by herbivores. This reduction is associated with reduced reproduction by caribou (Post et al. 2008).

In addition, climate warming appears to create asynchrony between the time of maximum phenological heterogeneity and timing of the peak of reproduction by temperate herbivores. In Greenland, caribou synchronize their reproduction with the onset and progression of plant growth, timing their spring migrations such that they arrive on calving grounds when plants are emerging from dormancy (Post et al. 2003). Spatial heterogeneity in plant maturity created by variation in elevation, slope, and aspect allows these animals to prolong the period of peak energy and nutrient intake as depicted in Fig. 6.2. However, climate warming causes earlier initiation of plant growth, which, in turn, creates a mismatch between caribou migration and plant phenology (Post and Forchhammer 2008, Post et al. 2008). This asynchrony compresses the time interval during which animals can obtain maximum intake rates, and, in so doing, harms reproductive success. Grave consequences could arise from the combined effects of fragmentation, which compresses the spatial scale of interaction between herbivores and plant communities and climate warming, which compresses the temporal scale of that interaction.

6.7   Conclusions: the importance of spatial context for population dynamics

Understanding the causes and consequences of temporal variation in factors shaping population dynamics has formed a dominant theme in population ecology for decades (Kingsland 1985). Efforts to understand the role of the spatial context, particularly as it interacts with time, form a much more recent endeavor. We reviewed findings to show that heterogeneity in resources over space may be as important as the total amount of resources in shaping dynamics of populations of mobile herbivores. The temporal trajectory of plant growth and senescence creates trade-offs between quality and quantity of forage resources for herbivores. These trade-offs have fundamental implications for herbivore condition and population performance. Moreover, heterogeneous landscapes may contain patches of resources that remain stable in time despite fluctuations in the average or total amount of the resource. In so doing, heterogeneity can stabilize dynamics of ungulate populations.

Much progress has been made in including realistic aspects of trophic interactions in models of herbivore population dynamics, progress that offers insight well beyond traditional formulations (Chapter 2). However, there is still a need for models that achieve theoretical generality while also including measurable state variables and parameters. Many of the models reviewed here depend on representing the biology of specific species, which limits their generality. The more general models (i.e., Owen-Smith 2002a, 2004) rely on a representation of heterogeneity in forage quality that cannot be measured. A promising approach to representing heterogeneity in a simple, yet observable way, is developing statistical distribution functions relating plant nutritional quality to biomass (Demment and Van Soest 1985, Hobbs and Swift 1985, Edelstein-Keshet 1986). Future modeling efforts might usefully exploit these distributions.

Acknowledgments

This work was supported by awards DEB0444711 (Effects of Habitat Fragmentation on Consumer-Resource Dynamics in Environments Varying in Space and Time) and DEB-0119618 (Biocomplexity, Spatial Scale and Fragmentation: Implications for Arid and Semi-Arid Ecosystems) from the United States National Science Foundation to Colorado State University and by the CSIRO Sustainable Ecosystems Division. Further support was provided in part by the National Science Foundation while Hobbs was serving as a rotating Program Director in the Division of Environmental Biology. Andrew Illius provided critical comments on a previous draft. Any opinions, findings, conclusions, or recommendations are those of the authors and do not necessarily reflect the views of the National Science Foundation.

References

Adler, P. B., D. A. Raff, and W. K. Lauenroth. 2001. The effect of grazing on the spatial heterogeneity of vegetation. *Oecologia* 128: 465–479.

Albon, S. D., and R. Langvatn. 1992. Plant phenology and the benefits of migration in a temperate ungulate. *Oikos* 65: 502–513.

Bailey, D. W., J. E. Gross, E. A. Laca, et al. 1996. Mechanisms that result in large herbivore grazing distribution patterns. *Journal of Range Management* 49: 386–400.

Baker, D. L., and N. T. Hobbs. 1982. Composition and quality of elk summer diets in Colorado. *Journal of Wildlife Management* 46: 694–703.

Bell, R. H. V. 1971. A grazing ecosystem in the Serengeti. *Scientific American* 225: 86–93.

Blatt, S. E., J. A. Janmaat, and R. Harmsen. 2001. Modelling succession to include a herbivore effect. *Ecological Modelling* 139: 123–136.

Boone, R. B. 2007. Effects of fragmentation on cattle in African savannas under variable precipitation. *Landscape Ecology* 22: 1355–1369.

Boone, R. B., S. B. BurnSilver, P. K. Thornton, J. S. Worden, and K. A. Galvin. 2005. Quantifying declines in livestock due to land subdivision. *Rangeland Ecology and Management* 58: 523–532.

Boone, R. B. and N. T. Hobbs. 2004. Lines around fragments: effects of fencing on large herbivores. *African Journal of Range and Forage Science* 21: 147–158.

Boone, R. B., S. J. Thirgood, and J. G. C. Hopcraft. 2006. Serengeti wildebeest migratory patterns modeled from rainfall and new vegetation growth. *Ecology* 87: 1987–1994.

Breman, H., and C. T. de Wit. 1983. Rangeland productivity and exploitation in the Sahel. *Science* 221: 1341–1347.

Caughley, G. 1976. Wildlife management and the dynamics of ungulate populations. In *Applied Biology*, ed. T. H. Coaker, vol. 1, pp. 183–244. Academic Press, London.

Caughley, G. 1982. Vegetation and the dynamics of modelled grazing systems. *Oecologia* 54: 309–312.

Caughley, G., G. C. Grigg, and L. Smith. 1985. The effect of drought on kangaroo populations. *Journal of Wildlife Management* 49: 679–685.

Caughley, G. and J. H. Lawton. 1981. Plant-herbivore systems. In *Theoretical Ecology*, ed. R. M. May. 2nd edition, pp. 132–166. Blackwell Scientific Pub, Oxford.

Coblentz, W. K., J. O. Fritz, W. H. Fick, R. C. Cochran, and J. E. Shirley. 1998. In situ dry matter, nitrogen and fiber degradation of alfalfa, red clover and eastern gamagrass at four maturities. *Journal of Dairy Science* 81: 150–161.

Colchero, F., R. A. Medellin, J. S. Clark, R. Lee, and G. G. Katul. 2009. Predicting population survival under future climate change: density dependence, drought and extraction in an insular bighorn sheep. *Journal of Animal Ecology* 78: 666–673.

Deinum, B., J. de Beyer, P. H. Nordfeldt, A. Kornher, O. Østgård, and G. van Bogaert. 1981. Quality of herbage at different latitudes. *Netherlands Journal of Agricultural Science* 29: 141–150.

Demment, M. W. and P. J. Van Soest. 1985. A nutritional explanation for body-size patterns of ruminant and nonruminant herbivores. *American Naturalist* 125: 641–672.

Doncaster, C. P. 2001. Healthy wrinkles for population dynamics: unevenly spread resources can support more users. *Journal of Animal Ecology* 70: 91–100.

Edelstein-Keshet, L. 1986. Mathematical theory for plant-herbivore systems. *Journal of Mathematical Biology* 24: 25–58.

Ellis, J. E. and D. M. Swift. 1988. Stability of African pastoral ecosystems: alternate paradigms and implications for development. *Journal of Range Management* 41: 450–459.

Ericsson, G., J. P. Ball, and K. Danell. 2002. Body mass of moose calves along an altitudinal gradient. *Journal of Wildlife Management* 66: 91–97.

Fabricius, C. 1994. The relation between herbivore density and relative resource density at the landscape level: Kudu in semi-arid savanna. *African Journal of Range and Forage Science* 11: 7–10.

Fahrig, L. 2002. Effect of habitat fragmentation on the extinction threshold: a synthesis. *Ecological Applications* 12: 346–353.

Fletcher, M., C. J. Southwell, N. W. Sheppard, et al. 1990. Kangaroo population trends in the Australian rangelands, 1980–87. *Search* 21: 28–29.

Forchhammer, M. C., E. Post, T. B. G. Berg, T. T. Hoye, and N. M. Schmidt. 2005. Local-scale and short-term herbivore-plant spatial dynamics reflect influences of large-scale climate. *Ecology* 86: 2644–2651.

Forchhammer, M. C., N. C. Stenseth, E. Post, and R. Langvatn. 1998. Population dynamics of Norwegian red deer: density-dependence and climatic variation. *Proceedings of the Royal Society of London Series B: Biological Sciences* 265: 341–350.

Frank, D. A. 1998. Ungulate regulation of ecosystem processes in Yellowstone National Park: direct and feedback effects. *Wildlife Society Bulletin* 26: 410–418.

Frank, D. A. and S. J. McNaughton. 1992. The ecology of plants, large mammalian herbivores, and drought in Yellowstone National Park. *Ecology* 73: 2043–2058.

Fryxell, J. M. 1991. Forage quality and aggregation by large herbivores. *American Naturalist* 138: 478–498.

Fryxell, J. M., J. Greever, and A. R. E. Sinclair. 1988. Why are migratory ungulates so abundant? *American Naturalist* 131: 781–798.

Fryxell, J. M., and A. R. E. Sinclair. 1988. Seasonal migration by white-eared kob in relation to resources. *African Journal of Ecology* 26: 17–31.

Fryxell, J. M., J. F. Wilmshurst, and A. R. E. Sinclair. 2004. Predictive models of movement by Serengeti grazers. *Ecology* 85: 2429–2435.

Fryxell, J. M., J. F. Wilmshurst, A. R. E. Sinclair, D. T. Haydon, R. D. Holt, and P. A. Abrams. 2005. Landscape scale, heterogeneity, and the viability of Serengeti grazers. *Ecology Letters* 8: 328–335.

Galvin, K. A., R. H. Behnke, N. T. Hobbs, and R. S. Reid (eds). 2008. *Fragmentation of Semi-Arid and Arid Landscapes: Consequences for Human and Natural Systems.* Kluwer Academic Publishers, Springer, Dordrecht.

Garrott, R. A., G. C. White, R. M. Bartmann, L. H. Carpenter, and A. W. Alldredge. 1987. Movements of female mule deer in northwest Colorado. *Journal of Wildlife Management* 51: 634–643.

Gordon, I. J. 1989. Vegetation community selection by ungulates on the Isle of Rhum. *Journal of Applied Ecology* 26: 35–51.

Gustavsson, A. M. and K. Martinsson. 2004. Seasonal variation in biochemical composition of cell walls, digestibility, morphology, growth and phenology in timothy. *European Journal of Agronomy* 20: 293–312.

Hassell, M. P. 1980. Some consequences of habitat heterogeneity for population-dynamics. *Oikos* 35: 150–160.

Hassell, M. P. and S. W. Pacala. 1990. Heterogeneity and the dynamics of host parasitoid interactions. *Philosophical Transactions of the Royal Society of London Series B: Biological Sciences* 330: 203–220.

Hebblewhite, M., E. Merrill, and G. McDermid. 2008. A multi-scale test of the forage maturation hypothesis in a partially migratory ungulate population. *Ecological Monographs* 78: 141–166.

Herfindal, I., E. J. Solberg, B. E. Saether, K. A. Hogda, and R. Andersen. 2006. Environmental phenology and geographical gradients in moose body mass. *Oecologia* 150: 213–224.

Hjeljord, O. and T. Histol. 1999. Range-body mass interactions of a northern ungulate–a test of hypothesis. *Oecologia* 119: 326–339.

Hobbs, N. T. 1989. Linking Energy Balance to Survival in Mule Deer: Development and Test of a Simulation Model. *Wildlife Monographs* 101.

Hobbs, N. T. 1999. Responses of large herbivores to spatial heterogeneity in ecosystems. In *Nutritional Ecology of Herbivores: Proceedings of the Vth International Symposium on the Nutrition of Herbivores*. H. G. Jung and G. C. Fahey, pp. 97–129. American Society of Animal Science, Savory, Illinois.

Hobbs, N. T. 2003. Challenges and opportunities for integrating ecological knowledge across scales. *Forest Ecology and Management* 181: 222–238.

Hobbs, N. T., K. A. Galvin, C. J. Stokes, et al. 2008. Fragmentation of rangelands: implications for humans, animals, and landscapes. *Global Environmental Change: Human and Policy Dimensions* 18: 776–785.

Hobbs, N. T., J. E. Gross, L. A. Shipley, D. E. Spalinger, and B. A. Wunder. 2003. Herbivore functional response in heterogeneous environments: a contest among models. *Ecology* 84: 666–681.

Hobbs, N. T. and D. M. Swift. 1985. Estimates of habitat carrying capacity incorporating explicit nutritional constraints. *Journal of Wildlife Management* 49: 814–822.

Holdo, R. M., R. D. Holt, and J. M. Fryxell. 2009. Opposing rainfall and plant nutritional gradients best explain the wildebeest migration in the Serengeti. *American Naturalist* 173: 431–445.

Holt, R. D. and M. P. Hassell. 1993. Environmental heterogeneity and the stability of host parasitoid interactions. *Journal of Animal Ecology* 62: 89–100.

Hume, D. E. 1991. Primary growth and quality characteristics of *Bromus willdenowii* and *Lolium multiflorum. Grass and Forage Science* 46: 313–324.

Illius, A. W. 2006. Linking functional response and foraging behaviour to population dynamics. In *Large Herbivore Ecology, Ecosystem Dynamics, and Conservation*, eds. K. Danell, R. Bergstrom, P. Duncan, and J. Pastor, pp. 71–96. Cambridge University Press, Cambridge, UK.

Illius, A. W. and T. G. O'Connor. 2000. Resource heterogeneity and ungulate population dynamics. *Oikos* 89: 283–294.

Jacobson, A. R., A. Provenzale, A. von Hardenberg, B. Bassano, and M. Festa-Bianchet. 2004. Climate forcing and density dependence in a mountain ungulate population. *Ecology* 85: 1598–1610.

Jefferies, R. L., D. R. Klien, and G. R. Shaver. 1994. Vertebrate herbivores and northern plant communities: reciprocal influences and responses. *Oikos* 71: 193–206.

Johnson, C. J., K. L. Parker, and D. C. Heard. 2001. Foraging across a variable landscape: behavioral decisions made by woodland caribou at multiple spatial scales. *Oecologia* 127: 590–602.

Kiker, G. 1998. *Analyses of Large Herbivore Distributions, Abundances, and Carrying Capacity Using Ecosystem Modeling*. PhD thesis, Cornell University, Ithaca, New York, USA.

Kingsland, S. E. 1985. *Modeling Nature: Episodes in the History of Population Ecology*. University of Chicago Press, Chicago, Illinois, USA.

Kolasa, J. and C. D. Rollo. 1991. Introduction: the heterogeneity of heterogeneity: a glossary. In *Ecological Heterogeneity*, eds. J. Kolasa and S. T. A. Pickett, pp. 1–23. Springer-Verlag, New York.

Laca, E. A. and M. W. Demment. 1991. Herbivory: the dilemma of foraging in a spatially heterogeneous food environment. In *Plant Defenses Against Mammalian Herbivory*, eds. R. T. Palo and C. T. Robbins, pp. 30–44. CRC Press, Boca Raton, Florida.

Langvatn, R., S. D. Albon, T. Burkey, and T. H. Clutton-Brock. 1996. Climate, plant phenology and variation in age of first reproduction in a temperate herbivore. *Journal of Animal Ecology* 65: 653–670.

Levins, R. 1966. The strategy of model building in population biology. *American Scientist* 54: 421–431.

Loft, E. R., R. C. Bertram, and D. L. Bowman. 1989. Migration patterns of mule deer in the central Sierra Nevada USA. *California Fish and Game* 75: 11–19.

McNaughton, S. J. 1979. Grazing as an optimization process–grass ungulate relationships in the Serengeti. *American Naturalist* 113: 691–703.

McNaughton, S. J. 1983. Serengeti grassland ecology: the role of composite environmental factors and contingency in community organization. *Ecological Monographs* 53: 291–320.

McNaughton, S. J. 1989. Interactions of plants of the field layer with large herbivores. *Symposium of the Zoological Society of London* 61: 15–29.

McNaughton, S. J. 1990. Mineral nutrition and seasonal movements of African migratory ungulates. *Nature* 345: 613–615.

Mduma, S. A. R., A. R. E. Sinclair, and R. Hilborn. 1999. Food regulates the Serengeti wildebeest: a 40-year record. *Journal of Animal Ecology* 68: 1101–1122.

Merrill, E. H. and M. S. Boyce. 1991. Summer range and elk population dynamics in Yellowstone National Park. In *The Greater Yellowstone Ecosystem*, eds. R. B. Keiter and M. S. Boyce. Yale University Press, New Haven, Connecticut, USA.

Mueller, T., K. A. Olson, T. K. Fuller, G. B. Schaller, M. G. Murray, and P. Leimgruber. 2008. In search of forage: predicting dynamic habitats of Mongolian gazelles using satellite-based estimates of vegetation productivity. *Journal of Applied Ecology* 45: 649–658.

Murray, M. G., and A. W. Illius. 2000. Vegetation modification and resource competition in grazing ungulates. *Oikos* 89: 501–508.

Mysterud, A., R. Langvatn, N. G. Yoccoz, and N. C. Stenseth. 2001a. Plant phenology, migration and geographical variation in body weight of a large herbivore: the effect of a variable topography. *Journal of Animal Ecology* 70: 915–923.

Mysterud, A., R. Langvatn, N. G. Yoccoz, and N. C. Stenseth. 2002. Large-scale habitat variability, delayed density effects and red deer populations in Norway. *Journal of Animal Ecology* 71: 569–580.

Mysterud, A. and E. Ostbye. 2006. *Effect of climate and density on individual and population growth of roe deer Capreolus capreolus at northern latitudes: the Lier valley, Norway.* *Wildlife Biology* 12: 321–329.

Mysterud, A., N. C. Stenseth, N. G. Yoccoz, R. Langvatn, and G. Steinheim. 2001b. Nonlinear effects of large-scale climatic variability on wild and domestic herbivores. *Nature* 410: 1096–1099.

Mysterud, A., N. G. Yoccoz, N. C. Stenseth, and R. Langvatn. 2000. Relationships between sex ratio, climate and density in red deer: the importance of spatial scale. *Journal of Animal Ecology* 69: 959–974.

Nordengren, C., A. Hofgaard, and J. P. Ball. 2003. Availability and quality of herbivore winter browse in relation to tree height and snow depth. *Annales Zoologici Fennici* 40: 305–314.

Noy-Meir, I. 1975. Stability of grazing systems: an application of predator-prey graphs. *Journal of Ecology* 63: 459–481.

O'Neill, R. V., D. L. DeAngelis, J. B. Waide, and T. F. H. Allen. 1986. *A Hierarchical Concept of Ecosystems.* Princeton University Princeton, New Jersey.

Owen-Smith, N. 2002a. A metaphysiological modelling approach to stability in herbivore-vegetation systems. *Ecological Modelling* 149: 153–178.

Owen-Smith, N. 2002b. *Adaptive Herbivore Ecology.* Cambridge University Press, Cambridge, UK.

Owen-Smith, N. 2004. Functional heterogeneity in resources within landscapes and herbivore population dynamics. *Landscape Ecology* 19: 761–771.

Pastor, J., Y. Cohen, and R. Moen. 1999. Generation of spatial patterns in boreal forest landscapes. *Ecosystems* 2: 439–450.

Pastor, J., B. Dewey, R. Moen, D. J. Mladenoff, M. White, and Y. Cohen. 1998. Spatial patterns in the moose-forest-soil ecosystem on Isle Royale, Michigan, USA. *Ecological Applications* 8: 411–424.

Pettorelli, N., S. Dray, J.-M. Gaillard, et al. 2003. Spatial variation in springtime food resources influences the winter body mass of roe deer fawns. *Oecologia* 137: 363–369.

Pettorelli, N., A. Mysterud, N. G. Yoccoz, R. Langvatn, and N. C. Stenseth. 2005. Importance of climatological downscaling and plant phenology for red deer in heterogeneous landscapes. *Proceedings of the Royal Society B: Biological Sciences* 272: 2357–2364.

Post, E., P. S. Boving, C. Pedersen, and M. A. MacArthur. 2003. Synchrony between caribou calving and plant phenology in depredated and non-depredated populations. *Canadian Journal of Zoology* 81: 1709–1714.

Post, E. and M. C. Forchhammer. 2008. Climate change reduces reproductive success of an Arctic herbivore through trophic mismatch. *Philosophical Transactions of the Royal Society B: Biological Sciences* 363: 2369–2375.

Post, E., C. Pedersen, C. C. Wilmers, and M. C. Forchhammer. 2008. Warming, plant phenology and the spatial dimension of trophic mismatch for large herbivores. *Proceedings of the Royal Society B: Biological Sciences* 275: 2005–2013.

Post, E. and N. C. Stenseth. 1999. Climatic variability, plant phenology, and northern ungulates. *Ecology* 80: 1322–1339.

Putman, R. J., J. Langbein, A. J. M. Hewison, and S. K. Sharma. 1996. Relative roles of density-dependent and density-independent factors in population dynamics of British deer. *Mammal Review* 26: 81–101.

Schmitz, O. J. 1993. Trophic exploitation in grassland food-chains: simple-models and a field experiment. *Oecologia* 93: 327–335.

Scholes, R. J. and S. R. Archer. 1997. Tree-grass interactions in savannas. *Annual Review of Ecology and Systematics* 28: 517–544.

Searle, K. R., N. T. Hobbs, and L. A. Shipley. 2005. Should I stay or should I go? Patch departure decisions by herbivores at multiple scales. *Oikos* 111: 417–424.

Searle, K. R., N. T. Hobbs, B. A. Wunder, and L. A. Shipley. 2006. Preference in patchy landscapes: the influence of scale-specific intake rates and variance in reward. *Behavioral Ecology* 17: 315–323.

Senft, R. L., M. B. Coughenour, D. W. Bailey, L. R. Rittenhouse, O. E. Sala, and D. M. Swift. 1987. Large herbivore foraging and ecological hierarchies. *Bioscience* 37: 789–799.

Shipley, L. A., A. W. Illius, K. Danell, N. T. Hobbs, and D. E. Spalinger. 1999. Predicting bite size selection of mammalian herbivores: a test of a general model of diet optimization. *Oikos* 84: 55–68.

Sinclair, A. R. E. and J. M. Fryxell. 1985. The Sahel of Africa–ecology of a disaster. *Canadian Journal of Zoology* 63: 987–994.

Underwood, N. 2004. Variance and skew of the distribution of plant quality influence herbivore population dynamics. *Ecology* 85: 686–693.

Van der Wal, R., N. Madan, S. van Lieshout, C. Dormann, R. Langvatn, and S. D. Albon. 2000. Trading forage quality for quantity? Plant phenology and patch choice by Svalbard reindeer. *Oecologia* 123: 108–115.

Wallmo, O. C., L. H. Carpenter, W. L. Regelin, R. B. Gill, and D. L. Baker. 1977. Evaluation of deer habitat on a nutritional basis. *Journal of Range Management* 30: 122–127.

Wang, G. M., N. T. Hobbs, R. B. Boone, et al. 2006. Spatial and temporal variability modify density dependence in populations of large herbivores. *Ecology* 87: 95–102.

Wang, G. M., N. Hobbs, S. Twombly, et al. 2009. Density dependence in northern ungulates: interactions with predation and resources. *Population Ecology* 51: 123–132.

Weisberg, P. J., N. T. Hobbs, J. E. Ellis, and M. B. Coughenour. 2002. An ecosystem approach to population management of ungulates. *Journal of Environmental Management* 65: 181–197.

Wilcove, D. S. and M. Wikelski. 2008. Going, going, gone: is animal migration disappearing? *Plos Biology* 6: 1361–1364.

Wilmshurst, J. F., J. M. Fryxell, B. P. Farm, A. R. E. Sinclair, and C. P. Henschel. 1999. Spatial distribution of Serengeti wildebeest in relation to resources. *Canadian Journal of Zoology* 77: 1223–1232.

7

Towards an ecology of population dynamics

Norman Owen-Smith

School of Animal, Plant and Environmental Sciences, University of the Witwatersrand, Johannesburg, South Africa

The working group "Dynamics of large herbivore populations in changing environments" that was established at the National Center for Environmental Analysis and Synthesis in 2001 had two basic aims:

1 To challenge prevalent population models using existing data on the dynamics of large herbivore populations
2 To develop alternative population models better able to accommodate the effects of environmental variability

My task in this final chapter is to outline the forms that these alternative models might take in the light of the factual information and concepts reviewed in the preceding chapters.

Weather influences have been revealed as pervasive, affecting forage production and its seasonal availability as well as imposing physiological stress (Chapter 3). Nevertheless, the magnitude of the mortality imposed depends on the population density, as well as predation and hunting pressure, influencing the effective resource availability. The birth mass and subsequent growth of the juvenile segment is most sensitive to resource limitations, affecting survival in this stage as well as the age at which females first reproduce (Chapter 4). Prime-aged females are most resistant to these influences, conferring population resilience, but with susceptibility to mortality decreasing with advancing senescence. Conditions in the year of birth may have a lasting effect on reproductive success at a

Dynamics of Large Herbivore Populations in Changing Environments, 1st edition. Edited by Norman Owen-Smith.
© 2010 Blackwell Publishing

later stage. Large herbivores respond to changes in vegetation induced by their consumption by switching to alternative resources, delaying or buffering the consequences of food depletion for irruptive dynamics (Chapter 5). Severe population crashes are almost always associated with extreme weather conditions, drastically reducing food availability in circumstances where populations have attained high abundance. Spatial heterogeneity in resource quality can prolong the period over which herbivores are able to exploit highly nutritious forage in the early growing season, promoting reproductive success (Chapter 6). Spatial heterogeneity in the amount of edible vegetation remaining during the dormant season may provide a buffer against high mortality in adverse years. Habitat fragmentation and barriers to movement limit opportunities to access such resource components, reducing population abundance and viability. How might this environmental complexity be represented in population models?

Large mammalian herbivores occupy a central position within food webs and hence present favorable opportunities to identify the dependency of population dynamics on changing vegetation resources as well as interactions with predators. Weather conditions influence the growth of plants constituting the food resource as well as the seasonal persistence and accessibility of this forage. Foraging responses can be documented in greater mechanistic detail than is possible for most insect, bird, and small mammal populations. Ungulates that have become domesticated provide more complete understanding of nutritional physiology than is available for most other organisms. Large ungulates are especially vulnerable to restrictions on the landscape scale over which they can move in seeking food and evading predation as a result of land-use changes induced by humans, affecting their capacity to cope with climate change.

The population models summarized in Chapter 2 generally place greater emphasis on endogenous population mechanisms than on ecological contexts influencing the population trends that emerge. It is axiomatic that density-dependent feedbacks on population growth rate must exist in order for populations to persist. The expected population growth must be positive at low abundance, and tend towards zero at some higher abundance level where resources and space become limiting. Exactly how the population growth rate changes between these density extremes depends on the processes operating. It is also obvious that population growth is the outcome of the survival and reproductive rates of the animals constituting the population, modified by movements between regions. However, these vital rates depend on the acquisition of the resources enabling survival as well as reproductive success, while also avoiding predation and excess physiological stress.

I approach modeling from the perspective of a field ecologist, acutely aware of the need for reliable theory synthesized into formal models, in order to address the practical problems raised by herbivore populations that are increasing excessively or declining towards unviable numbers. From the suite of models outlined in Chapter 2, I seek those that will be most pragmatically useful in transforming information into understanding, thereby providing a sound basis for decision-making. Some models may be dazzling in their intellectual sophistry, but prove unhelpful when applied to real-world situations. Here are some examples of conservation issues where models are needed because experimentation is difficult or impractical:

- Can widespread declines in ungulate populations within protected areas in Africa (Ogutu and Owen-Smith 2003, Caro and Scholte 2007) be ascribed to illegal hunting, land-use changes, or emerging consequences of climate change?
- What is the likely trajectory of elephant populations, vegetation features, and biodiversity in protected areas following the suspension of culling (Owen-Smith et al. 2006)?
- How much difference will the introduction of wolves make to the abundance of trees, elk, and other ungulates in Yellowstone National Park (Smith et al. 2003)?
- What is the capacity of forest remnants in Europe to support re-introduced bison (J. Cromsigt, personal communication)?

All models entail simplification, and a fundamental issue is what to leave out and what to retain. Whatever might be desirable, the available information limits what is practically achievable. Hence rather than advocating any one approach, I will suggest how environmental influences might be incorporated into models ranging from phenomenological descriptors to representations of individually responding agents.

7.1 Phenomenological descriptors

The simplest models express population growth as negatively dependent on increasing population size or density, incorporating two parameters that control respectively (i) the maximum growth rate towards low density, and (ii) the upper limit to the density level attained, or slope of the decline towards this density (eqn. (2.1–2.6)). Drawing on dynamical systems theory, the zero-growth level is interpreted as an equilibrial attractor in state space, subject to stochastic perturbations from exogenous factors and hence manifested as a stationary abundance distribution around some mean (Royama 1992, Dennis and Taper 1994, Bjornstad and Grenfell 2001, Coulson et al. 2004a). Royama (1992) further distinguished vertical perturbations of the density-dependent response, affecting the maximum increase rate as well as the equilibrium level, from lateral perturbations affecting only the equilibrium density (see Fig. 2.2).

However, the carrying capacity symbolized as K in the logistic equation is not stationary when related to the vegetation components constituting the food resource for herbivores. In African savannas, annual plant growth depends on rainfall, which varies widely between years (Rutherford 1980). Accordingly, the abundance attained by savanna herbivores, both in aggregate (Fritz and Duncan 1994) and individually (East 1984), is governed fundamentally by the annual rainfall total. The amount of herbage remaining green is affected additionally by rainfall received during the normally dry season, determining food quality during this bottleneck period. In temperate latitudes, plant growth depends on the timing of initiation dependent on spring temperatures, and on summer rainfall sustaining this growth, but with resultant forage production perhaps less variable between years than in rainfall-driven savannas. In regions with substantial winter snow, food availability during winter depends on snow depth and formation of ice crusts restricting access to the herbage beneath. The availability of food to herbivores is modified additionally by the population demand relative to this supply, i.e. on intraspecific competition, which effectively takes the form of a scramble.

As recognized by Caughley (1976) and Berryman et al. (1995), the logistic growth equation crudely expresses the dependence of population growth on the effective food share per individual, i.e. the ratio between supply and demand:

$$r_t = r_{max}\,[1 - (N_t/K_t)] \tag{7.1}$$

where N_t is the population size or density at time t, r_t is the relative population growth rate between times t and $t + 1$, r_{max} is the maximum growth rate towards low abundance, and K_t is the abundance level at which the population growth rate reaches zero, noted as being time-dependent. Hence the logistic equation can be interpreted as capturing resource-dependent population growth, with the density level modifying effective resource availability per capita. Arditi and Ginzburg (1989) and Berryman et al. (1995) advocated this resource ratio-dependent perspective.

Strictly, the value of K relative to r_{max} simply determines the slope of the decline in growth rate with increasing density, if this is indeed linear. Because the net growth rate represents the difference between population birth and death rates (excluding emigration and immigration), any factor influencing survival or reproduction will affect this slope, depending on how its effect is modified by changing density. Hence it is misleading to imagine K as a constant carrying capacity. Nevertheless, the concept of some upper limit to the population abundance that can be supported by resources may be useful, allowing that this ceiling level is likely to vary spatially and temporally.

Equation 7.1 implies that the population growth rate declines with increasing density even at low abundance levels where food may not be limiting. This can be corrected by assuming that food limitation takes hold only above some threshold density level N^*:

$$r_t = r_{max} \left[1 - \left\{ (N_t - N^*) / (K_t - N^*) \right\} \right], \text{ if } N_t > N^*$$

$$r_t = r_{max}, \text{ if } N_t \leq N^* \tag{7.2}$$

generating a plateau and ramp relationship (see Fig. 2.1e). This type of formulation was used to represent the dynamics of Soay sheep by Stenseth et al. (2004), assuming that adverse weather accentuated the density feedback causing the population to decline (see Fig. 3.1a).

Since the breakpoint may not be sharp, it can be more appropriate to represent the overall form of the density response as convexly curvilinear, by incorporating a θ-coefficient into the logistic equation as a shape parameter (following Fowler 1981; see eqn. (2.2)). This can be misleading if the consequently accelerating reduction in the population growth rate is extrapolated much beyond the zero-growth level. Severe early mortality reduces the demand on resources, alleviating food stress for remaining animals (Hallett et al. 2004), so that an inflection point might be expected towards very high abundance. Following Getz (1996), this pattern can be represented by plotting the multiplicative growth rate $(1 + r_t)$ against population abundance using the generalized sigmoid growth equation (eqn. (2.7); Fig. 7.1a). Accordingly, three distinct regions of the density response may be distinguished: (i) the low-density region where resources are not limiting, (ii) the intermediate region where population growth rate declines as resource limitations increasingly affect survival and reproduction, and (iii) a very high-density region where the feedback attenuates because of compensatory food release to the survivors (Fig. 7.1b). The point of inflection means that the density response may appear convex in its initial segment but concave over some higher abundance region, with the actual location of the zero-growth level dependent on circumstances.

The slope of the population growth curve in the neighborhood of the zero-growth region largely determines the propensity for oscillations to be generated, especially in response to environmental variation affecting the carrying capacity (Fig. 7.2). The simple theta-logistic equation incorporating a subtraction term tends to generate negative numbers for even

(a)

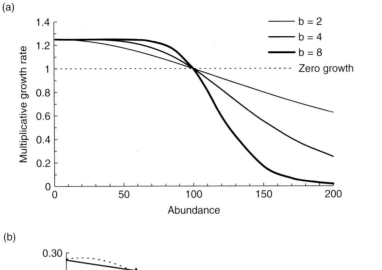

(b)

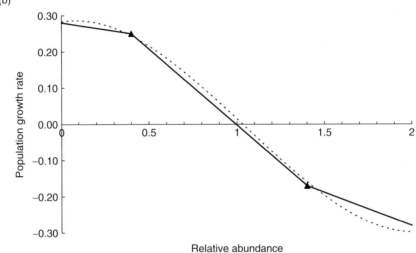

Figure 7.1 (a) Multiplicative population growth rate over a wide density range calculated from the generalized hyperbolic $N_{t+1}/N_t = (1 + r_t)/\{1 + (N_t/K_t)^b\}$ for different values of b (following Getz 1996). (b) Piecewise linear approximation of a hypothetical density response distinguishing three segments: (i) truncation of population growth rate by additive mortality towards low abundance, (ii) progressive decline in population growth rate through the zero-growth region, and (iii) reduced response at higher abundance levels due to individual differences and progressive mortality. *Dotted line* indicates overall curvilinear fit.

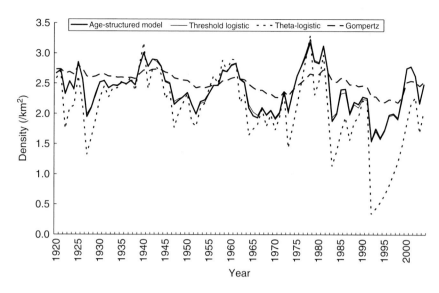

Figure 7.2 Population dynamics generated by alternative models with K carrying capacity dependent on annual rainfall totals recorded in Kruger Park: (i) age-structured kudu model, (ii) threshold Ricker model ($r_0 = 0.22$, $K = 2.75$ animals per km², threshold $N^*(t) = K^*(t)/3$), (iii) Ricker model incorporating θ−coefficient ($r_0 = 0.22$, $K = 2.75$ animals per km², $\theta = 3$), and (iv) threshold Gompertz model ($r_0 = 0.22$, $K = 2.75$ animals per km², threshold $N^*(t) = K^*(t)/3$).

moderate values of θ if the effect of rainfall variation on K is sufficiently great.

Predation can modify the density response, depending on how it interacts with food limitation. Strictly additive mortality across all age classes, as may be imposed by ambush predators (or human hunters), depresses both the zero-growth level and the potential population growth rate at low abundance (a vertical shift from Royama's perspective; Fig. 7.3a). However, if coursing predators merely amplify susceptibility to mortality from malnutrition, only the zero-growth level and not the recovery potential of the population is altered (Royama's lateral shift; Fig. 7.3b). These are idealized abstractions and real-world patterns will be less simple. The effect of predation may also vary over different ranges in prey density.

7.2   Time series elaborations

Besides lagged density feedbacks, time series models can potentially incorporate other influences on population change acting directly or with some

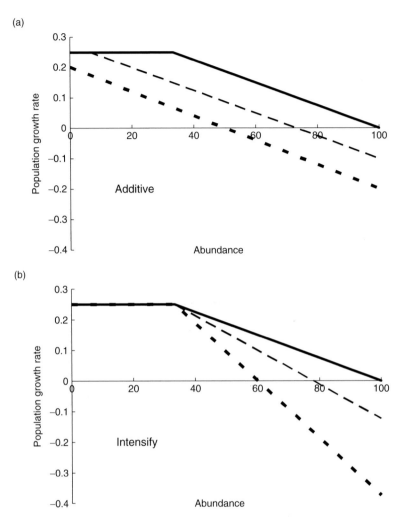

Figure 7.3 Alternative representations of the effects of predation on the density-dependent response, comparing neutral (*solid line*) with moderate (*dashed line*) and severe (*dotted line*) predation pressure. (a) Solely additive effect; (b) Intensification of density feedback.

time delay. Delayed effects of changing density are presumably generated by interactions with predators or resources, not represented directly when information on these other populations is lacking. Weather influences are generally interpreted as exogenous perturbations of the trophic interactions (Turchin 2003).

For large herbivores, the delays associated with trophic interactions are likely to extend over several years, because of the generation times of both large mammal and plant populations (Lande et al. 2002). Moreover, the more immediate effects of the prevailing density operate over some diffuse period rather than from the density level one annual time step earlier. Population density relative to resources affects fertility, growth of the fetus, neonatal survival, and subsequent survival of offspring plus older animals until the time of the census. A lagged feedback from density 2 years earlier can be associated with effects on fecundity (Forchhammer et al. 1998), while the density level 1 year earlier influences fetal growth. The density prevailing over the annual cycle leading up to the time of the census affects survival rates through the adverse season. In practice, the density levels of ungulate populations differ little between successive years, because generally 80% or more of the population survives from one year to the next unless an unusual die-off occurs. Differences in the abundance estimates between successive years are due mainly to census errors, potentially generating spurious indications of density dependence in annual population growth rates calculated as the difference between these estimates (Solow 2001). An averaged abundance estimate suppressing the negative autocorrelation in successive estimates of population change may be more revealing of the actual density effect (Owen-Smith and Mills 2006).

If weather conditions perturb population growth additively, they can be included as additional predictors in a time series model:

$$r_t = r_{max} + b_0 N_\tau + b_1 W_{1,t} + b_2 W_{2,t} + \ldots \ldots \tag{7.3}$$

where τ represents the period over which the direct density feedback operates, $W_{x,t}$ represents one or more weather covariates, and the bs are fitted coefficients. Because r_{max} is the maximum population growth rate, these coefficients must all be negative. Nonlinear relationships can be accommodated through suitable transformations of the predictors, ideally after they have been normalized relative to their mean values.

If adverse weather conditions affect the susceptibility of animals to mortality predisposed by food shortages, without changing the food supply, the threshold level for the onset of density dependence should remain constant while the slope of the density response beyond the threshold steepens (assuming a linear form):

$$r_t = r_{max}, \text{ if } N_t \leq N^*$$
$$r_t = r_{max} - bW_t(N_t - N^*), \text{ if } N_t > N^* \tag{7.4}$$

where N^* represents the threshold density. However, if weather primarily influences vegetation growth and hence food production for herbivores, a resource-ratio representation becomes appropriate:

$$r_t = r_{max}, \text{ if } N_t \leq N^*$$

$$r_t = r_{max} - b(N_t - N^*)/W_{R,t}, \text{ if } N_t > N^* \tag{7.5}$$

where the weather component represented by rainfall is incorporated inversely because its direct influence on food production is positive. The location of the breakpoint N^* will also shift if it retains a constant *proportional* relationship with the zero-growth level.

A complication arises when the interacting variables require different transformations, e.g. rainfall effects acting proportionately via a log-transformation, with those of density effectively linear on an arithmetic scale. This distinction may justify retaining the separate additive contributions of these factors as well as some measure of the interaction.

Incorporating multiple weather components separately dilutes the capacity of the data to distinguish their relative influences, especially when they are not independent. In these circumstances it becomes beneficial to use aggregate measures like the North Atlantic Oscillation, indexing the association between temperature and precipitation conditions experienced during winter in high northern latitudes (Stenseth et al. 2003). Because winter weather leads to deaths by draining energy reserves, the effect is additive, contrary to the influence of El Nino Southern Oscillation (ENSO) on food production for tropical ungulates.

Weather influences may also be delayed in their expression. For example, prior rainfall conditions may have a lasting effect on vegetation structure or composition, and hence affect the resource capacity supporting herbivores independently of the current rainfall (Owen-Smith and Mills 2006). A lagged influence may also arise through cohort effects, whereby adverse conditions in the year of birth can have a persistent effect on subsequent reproductive performance (Chapter 4). It is preferable to express the relationships with specific causal processes directly rather than indirectly through lagged density feedbacks.

7.3 Environmental structure

I will now make a radical departure from convention by turning my attention next to environmental structure, the constituents of K carrying capacity, before considering population structure, the components of abundance N. The above models simply represented annual variation in density, rainfall, and other weather aspects. However, weather conditions and hence resource availability generally vary widely over the seasonal cycle, and the extreme conditions may have a greater influence on the annual population dynamics than aggregate measures.

The model scaffolding is provided by coupled consumer–resource equations, originally formulated to represent predator–prey oscillations over multiyear periods. My elaboration builds on Caughley's (1976) basic herbivore–vegetation model (see also Forsyth and Caley 2006), developing a more explicitly metaphysiological representation of the trophic interactions (see Getz 1993, 1999, Owen-Smith 2002a, 2005). To incorporate these material fluxes, population abundance must be expressed in units of biomass density, rather than numerical density as is conventional. The biomass gain is dependent on the food intake rate in relation to changing food availability (or "functional response"), taking into account the nutritional value of the food consumed. Biomass losses need to be partitioned between physiological costs and distinctive sources of mortality:

$$(1/H)dH/dt = G - M_P - M_Q - M_C \qquad (7.6)$$

where H = herbivore biomass density, G = rate of biomass gain from food consumed, M_P = rate of metabolic attrition, M_Q = mortality loss dependent on nutritional gains relative to requirements, and M_C = mortality loss to predation or other causes independent of nutritional status (Fig. 7.4). This has been labeled the **G**rowth, **M**etabolism and **M**ortality, or GMM, model (Owen-Smith 2002a). Note the differential equation formulation, indicating that biomass fluxes are effectively continuous over time steps of a day or longer, in contrast to numerical population changes concentrated around a birth pulse. The rate of biomass gain G is a function of food availability, e.g., using the Michaelis-Menton equation (or Holling Type II

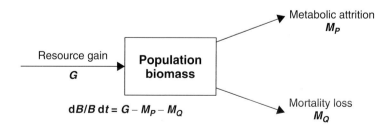

Figure 7.4 Schematic outline of the Growth, Metabolism, and Mortality (or GMM) model.

functional response) as an approximation,

$$G = ci_{max}F/(f_{1/2} + F) \tag{7.7}$$

where F = available food biomass, i_{max} = maximum rate of food intake, $f_{1/2}$ = amount of food at which the intake rate reaches half of its maximum, and c = conversion coefficient from food consumed into herbivore biomass. A piecewise linear function, reflecting the upper limit to the food intake rate, could alternatively be used. Although interference competition is not represented, a density feedback arises indirectly through the impact of consumption on subsequent resource availability. The lag time involved is effectively zero, i.e. less food is available tomorrow if more is eaten today.

When food is superabundant (i.e. $F \gg f_{1/2}$), the population will increase at its maximum rate $ci_{max} - M_P - m_{min}$, where the minimum mortality rate m_{min} includes mostly terminal mortality due to senescence. This maximum rate of biomass increase must equal the maximum population growth rate r_{max}, allowing for restrictions on the growth potential of different population segments. Accordingly, r_{max} is determined basically by the maximum rate at which food can be consumed, the nutritional value of this food, basic metabolic costs, and the potential life span. The rate at which immature animals can grow in mass as well as reproductive investments is constrained by the interplay between features of the environment and basic animal physiology.

With diminishing food availability, mortality losses will rise as a consequence of nutritional deficiencies. Assuming an inverse linear relationship,

$$M_Q = m_{min} + m(M_P/G) \tag{7.8}$$

where m = slope coefficient governing the magnitude of the rise in mortality with increasing food deficiencies relative to metabolic requirements.

An anomaly arises with respect to time frames if the food intake rate underlying G is measured on a daily time scale, whereas mortality losses depend on food gains integrated over some period back into the past. This can be surmounted by loss accounting over discrete time periods. The maximum population growth potential is reduced through various sources of mortality and other losses as well as through metabolic attrition. Most basically,

$$(1/H)\Delta H/\Delta t = r_{max} - m(M_P/G) \tag{7.9}$$

assuming that resource limitations are expressed solely through the resultant mortality loss M_Q, dependent on resource gains relative to metabolic demands over some period back in time, and that there is no additive mortality due to predation or hunting. Specifically, past resource gains in excess of immediate requirements can be stored and carried over to subsidize deficiencies arising at a later time. More generally, reductions in the population growth potential arise through several mechanisms: failure to conceive; failure to carry the fetus to term; neonatal mortality; mortality from various causes in later stages – all dependent fundamentally on limitations in the resource supply relative to demands. The maximum potential rate of biomass gain r_{max} is a species-specific attribute governed by physiological features controlling resource capture and conversion, the basic metabolic rate, and life history, these features presumably being adapted to the environmental capacity to support this rate of gain in biomass. I am ignoring the specific dependency of potential gains and losses on the age, size, or life history stage of the individuals constituting the population, assuming that the population biomass is made up by some fixed mix of these individual states, which is not strictly true. This assumption will be relaxed later.

The structural similarity between eqns (7.9) and (7.1) is evident. The nebulous carrying capacity K has been replaced by a measure of the rate of biomass gain G, while the density measure is now expressed in terms of the metabolic demand associated with the population biomass. The proportional growth in biomass over the annual cycle is the product of the relative population growth over each of the discrete stages of the year distinguished. The general pattern is for herbivore biomass to increase over the growing season when food is abundant and relatively high in quality, via the growth in individual mass especially of juvenile recruits, and shrink during the dormant season when animals lose body condition and become subject to mortality as a result (Fig. 7.5). Little or no biomass increase is associated with the birth pulse, which entails merely a subdivision of maternal body mass between the mother and the offspring. As a minimum, two divisions of the year should be distinguished, but further subdivisions may be more revealing. Seasonal distinctions in population density and other feedbacks on population growth have been modeled for small mammals by Stenseth (1999; see also Stenseth et al. 2002).

Having accommodated temporal variation over the annual cycle, we next recognize that herbivores generally depend on distinct components of the vegetation resource at different stages of the year, especially when seasonal movements occur. Hence the aggregate food resource F needs to be partitioned among its components F_1, F_2, F_3.... During the early growing season, herbivores seek areas where young vegetation provides highly nutritious forage to support the growth to term of the fetus and subsequent demands of lactation. Later in the growing season, herbivores

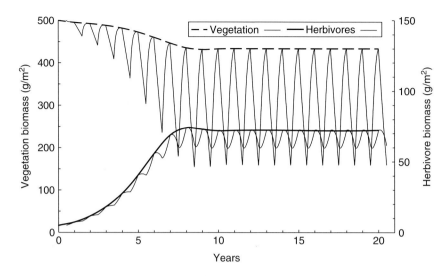

Figure 7.5 Output of the GMM model showing annual oscillations in herbivore and vegetation biomass through seasonal cycles in relation to smoothed lines obtained from once-annual estimates of herbivore and vegetation abundance conducted at the end of the growing season when both are at their peak values.

fall back on plants that remain generally high in nutritional value, with resource depression through consumption being inconsequential as long as regrowth occurs so that forage remains abundant. With the onset of the dormant season, herbivores seek out places where forage of adequate nutritional value remains, with the resource supply diminishing as plants senesce and remaining forage is consumed. By late winter or the dry season, herbivores fall back on whatever forage remains, relying on fat reserves to counteract nutritional deficiencies. Vegetation components supporting herbivores through these stages can be distinguished functionally as high-quality (or *"kick-start"*) resources exploited early in the growing season, *staple* resources utilized later in the growing season, *reserve* resources turned to early in the dormant season, and *buffer* resources retarding starvation late in the dormant season (Owen-Smith 2002a). Accordingly, at least four seasonal divisions need to be distinguished from this perspective. Additional subdivisions may be required to represent the resources bridging critical periods, e.g. the time between the end of the dry season and start of the wet season in years when rain is delayed. Awareness of this resource structure highlights the need for herbivores to obtain adequate resources through each stage of the seasonal cycle (interpreted as "stepping stones" by Owen-Smith and Cooper 1989) in order to maintain a population.

Thus far I have considered the response of the herbivores to changes in the quantity and quality of their food resource over the seasonal cycle, potentially extended through multiple years. Changes in the vegetation components generating this supply can be represented by subtracting the food intake by the herbivores from the inherent growth potential of the vegetation:

$$dV/dt = r_{v,t}V - i_{max}FH/(f_{1/2} + F) \qquad (7.10)$$

where $r_{v,t}$ represents the biomass growth function of the vegetation V and the term on the right represents the rate of consumption of the edible fraction F by the herbivores, modified from eqn. (7.7). However, immediately we have two mismatches. Vegetation growth is seasonally phased, rather than taking place continuously. Most vegetation growth takes place early in the growing season, much faster than herbivores can consume it, with subsequent regrowth partially or completely replacing what the herbivores have eaten until the dormant season ensues. Vegetation then becomes a nonrenewing resource, with much of the edible biomass becoming converted to necromass in the form of fallen leaves and brown grass. Furthermore, the edible vegetation F represents a small fraction of the total vegetation V generating growth. Specifically, much of the plant biomass is under ground and hence out of reach to ungulates.

In practice, large herbivores experience little change in their rate of food intake in response to changing food abundance over a wide range, so that factors controlling bite size and biting rate become the main constraint (effectively determining i_{max} in eqn. (7.10); Spalinger and Hobbs 1992). Changes in the nutritional value of the material consumed have the major influence on biomass gains, with nutrient contents within plant parts changing widely over the seasonal cycle (Chapter 6). Depletion of preferred vegetation components can occur at times, e.g. when grazing lawn grasslands are grazed down to a height where the rate of gain becomes inadequate to meet metabolic needs (Murray and Brown 1993), but with the potential for less nutritious plant types to be substituted. For both wildebeest (Wilmshurst et al. 1999) and gazelles (Fryxell et al. 2005) on the Serengeti plains, animal movements were governed more by the local greenness of the grass on offer than by changes in grass height and hence biomass. Shifts between vegetation components in response to depletion consequently lead to reductions in diet quality. While measuring the changing quantity of multiple vegetation components could be an overwhelming challenge, assessing changes in forage and diet quality is more easily accomplished, by chemically analyzing the herbage or the fecal residues after digestion.

Shifts in the dietary dependency of herbivores may take place within a fine-scale mosaic, or by movement or migration over wider scales within regional landscapes. Greater movement entails higher metabolic costs, hence raising the magnitude of the term M_P relative to nutritional gains. Changes in metabolic expenditures arising through movements could be monitored using global positioning system (GPS) tracking technology. Depletion of local food resources within foraging areas, or a drop in their nutritional value, prompts relocation to better foraging conditions elsewhere. Documenting the resultant resource utilization distribution (Kernohan et al. 2001), controlled for activity performed, indicates the effective resource yields obtained from different regions of the available landscape. Prevailing weather conditions also influence the effective value of M_P, through imposing a metabolic drain or by restricting foraging time. How to represent extreme weather events occurring as brief pulses within a season remains a challenge, unless time is resolved to a daily or at least weekly step. Stored body reserves could be represented by partitioning population biomass between a constitutive component, represented by lean body mass, and a labile component, represented by fat deposits carried through from flush times to bridge seasonal shortfalls in food gains (Getz and Owen-Smith 1999). The dynamic energy balance approach advocated by De Roos and others (De Roos and Persson 2001, De Roos et al. 2009) emphasizes physiological processes in greater detail, but gives less attention to environmental structure.

Sections of the landscape drawing herbivore concentrations at particular stages of the seasonal cycle have been interpreted as nutritional "hotspots" (McNaughton 1988, Hebblewhite et al. 2008), or as "key resource areas" sought out during the dry season (Scoones 1995, Illius and O'Connor 1999). A metaphysiological modeling application indicated how disrupting herbivore migrations in Serengeti through confining subpopulations to solely the hot-spot constituted by the short-grass plains, or only the key resource area represented by the high-rainfall grasslands in the north, would drastically alter the herbivore abundance supported and dynamics manifested (Owen-Smith 2004). Besides food resources, access to surface water for drinking is an additional constraint on the exploitation of different landscape regions by grazing herbivores during the dry season, leading to local resource depletion in proximity to perennial water sources (Andrew 1988). Over-provision of waterholes and consequent elimination of buffer resources remote from water to bridge the transition from prolonged drought conditions to eventual arrival of the rains led to extreme population crashes of several herbivores species in a private nature reserve near Kruger Park (Walker et al. 1987).

A finer level representation of vegetation resources could be modeled, if sufficient information is available, with herbivores shifting their food selection among these plant types and parts in response to patterns of growth and depletion. This requires an algorithm to represent the adaptive response of the herbivores in terms of changing diet selection. One of the two alternatives may be invoked (Chapter 5 in Owen-Smith 2002a, Fryxell et al. 2004):

1 Concentration solely on the food types maximizing net gains during each time step, which is adequate if the temporal resolution is sufficiently fine to capture the resultant distribution of consumption over vegetation components.
2 Distributing consumption over food types in relation to their effective value (a matching response), which may more realistically represent situations where herbivores have imperfect information about the relative value of alternative resources, and restrictions on their movements.

Whether food types are intermingled, or found in different localities, determines whether dietary expansion and contraction, or food switching, occurs when availability changes.

A metaphysiological model assessing the population performance of grazing herbivores based on the composition of the grass layer was formulated in Chapter 11 of Owen-Smith (2002a). This chapter also developed a habitat suitability model for a browsing ungulate, which successfully predicted observed population abundance levels (the effective value of K carrying capacity) over a wide range using realistic parameter values (Fig. 7.6). At a more abstract level, this modeling approach demonstrated how functional heterogeneity in food resources could dampen the propensity of large herbivore populations to oscillate in abundance (Chapter 13 in Owen-Smith 2002a, b). In particular, such models expose the importance of buffer resources, intrinsically of low nutritional value but slowing rates of starvation during critical periods (see also Chapter 6 of the current book). The fallen leaf litter exploited by white-tailed deer on Anticosti Island during winter (Tremblay et al. 2005) may play this role in suppressing the population crash expected as a result of the high herbivore density attained. Restrictions on herbivore movements limit opportunities to exploit this heterogeneity, accentuating density feedbacks and leading to lowered herbivore abundance (Wang et al. 2006).

A crucial aspect of landscape heterogeneity not covered in Chapter 6 is spatial variation in the risk of predation, dependent on habitat structure. Thick bush or tall grass can greatly hamper the ability of herbivores to detect an approaching predator at a safe distance, and also hinder evasion

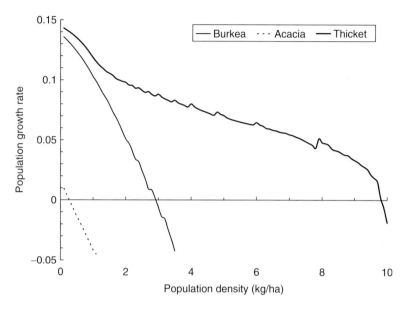

Figure 7.6 Output of a GMM habitat suitability model projecting the zero-growth density for kudu in three vegetation types. Highest density is projected for Thicket because the largely evergreen shrubs retain foliage through the dry season, intermediate density for Burkea savanna where trees are mostly deciduous, and very low density for Acacia savanna where trees retain almost no foliage by the late dry season. These projected densities agree closely with observed kudu densities in these habitats.

for those species relying on speed to escape. In northern latitudes, deep snow hinders ungulate movements and hence escape from wolves (Mech et al. 1987). Mountain ungulates occupy localities within or close to escape terrain in the form of steep slopes that they are better adapted to traverse than predators (Festa-Bianchet 1988). Hence habitat use and resultant nutritional gains depend not only on the resources on offer within these habitats, but also on the associated security from predation. These considerations relate to effective value of the additive mortality loss M_C in eqn. (7.6). Representing it spatially defines the "landscape of fear" (Brown et al. 1999), reflecting the likelihood of encountering a predator and relative likelihood of a successful attack in different regions. Within Kruger Park, artificially augmenting water points widened the distribution of the principal prey species for lions and hence the presence of these predators, occluding the refuge areas of low predation risk that rarer antelope species occupied within the park (Owen-Smith and Mills 2006). As a result, drastic reductions in populations of these antelope species occurred when lions switched their prey selection towards them after their main prey species became less readily available (Owen-Smith and Mills 2008).

The concept of source–sink areas or habitats (Pulliam 1988) may also become relevant at the landscape scale, bringing into consideration the contribution of dispersal movements to longer term population dynamics. Under constant environmental conditions, these movements would be expected to produce an "ideal free distribution," if less suitable conditions in the sink are counterbalanced by less crowding. However, if sink habitats fluctuate widely in the resources they provide or predation risk they confer, they may at times become more favorable than the source habitat, resulting in colonization waves followed by population attenuation. Local disparities in survival and recruitment have been documented for Soay sheep on Hirta, related to regional variation in grass quality, with dispersal between regions preventing local population divergence (Coulson et al. 1999). Regional variation in the increase of white rhinos within the Hluhluwe-Imfolozi Park indicated ongoing dispersal from the high density core into outlying areas, primarily by the subadult segment of both sexes (Table 11.9 in Owen-Smith 1988). Spatial disparities in population growth have also been documented for elephants in Zimbabwe (Chamaille-Jammes et al. 2008). Hence eqn. (7.6) would need to be expanded to accommodate shifts in the biomass distribution between regions as a result of such dispersal movements.

Projecting the likely responses of herbivores to changing circumstances through time, taking into account uncertainty, and tradeoffs between food gains and predation risk, requires a dynamic approach to optimization analysis (Clark and Mangel 2000). This procedure has been difficult to apply when more than three or four choices must be considered, because of the multiplying dimensionality introduced. However, by dividing the year into a sequence of stages, and limiting habitat choices within each, such models could be made tractable. These models would represent the outcome at the population level of the aggregate choices made by the animals constituting the population biomass. Formulating and testing such models remains a future challenge.

7.4 Population structure

Classical matrix models project the population growth rates generated by particular combinations of vital rates, assuming that these rates remain constant, i.e. uninfluenced by changing density or environmental conditions. Retrospective analysis assesses the consequences of observed variation in these rates for population growth, but not the sources of this variability (Caswell 2001). Findings have supported the generalization that juvenile survival is more sensitive to variation in density and weather conditions than survival rates among older animals (Eberhardt 2002). Many studies have recorded how the survival rates of juveniles and adults respond to density and weather variation, as documented in Chapter 3. While most of the annual variation in numerical population size or density is an outcome of variable juvenile survival (Gaillard et al. 1998), this is less true for population biomass dynamics, because new recruits constitute a relatively smaller proportion of this biomass than adults. When population trend has changed substantially, this has been due largely to reductions in adult survival, exerting much greater leverage over the population growth rate in both numerical and biomass terms (e.g. red deer, Coulson et al. 2004b; kudu and other African ungulates, Owen-Smith and Mason 2005). The nonlinear shape of the overall population growth response to changing density is governed partly by the different density regions over which particular demographic stages become affected by resource limitations (Owen-Smith 2006).

Demographically structured models accommodating environmental variability were developed for Soay sheep on Hirta (Coulson et al. 2001, 2008), kudu in Kruger Park (Owen-Smith 1990, 2000), and elk in Yellowstone (Wilmers and Getz 2004). These models also incorporated the increasing susceptibility to mortality associated with advancing senescence, revealing how changing age structure as well as shifting vital rates contributed to changing population trends (Festa-Bianchet et al. 2003, Coulson et al. 2005, Owen-Smith 2006).

However, for most large herbivore populations, information on the changing age structure is not available at this level of resolution. At best, juveniles born during the past year, and perhaps also yearlings, may be distinguished from older animals within the female segment. For males, it may be possible to distinguish a wider range of year classes from distinctions in horn or antler development.

Coulson et al. (2008) outlined a potential unification between demographically structured and unstructured time series models of population growth, using Soay sheep as the example. This entailed separating the recruitment and survival contributions in the time series formulation. Partitioning of the population biomass among growth stages could also be represented in a metaphysiological model, recognizing distinctions in metabolic rates and biomass growth potential. A population composed entirely of fully grown males would have no biomass growth potential, but one made up of young males could increase in biomass through the growth of these individuals. The growth potential of adult females is expressed through their contribution to the growth of offspring, both in the fetal stage and after birth through provisioning via milk.

A stage-structured metaphysiological model for Soay sheep was formulated in Chapter 13 of Owen-Smith (2002a). This distinguished four growth stages: (i) juveniles, with their growth subsidized maternally prior to weaning, (ii) still-growing immature animals of both sexes, (iii) adult females diverting their growth potential to support the growth of dependent offspring, and (iv) adult males with no growth potential apart from changes in fat reserves. Biomass changes in these segments were simulated using a weekly iteration and a heterogeneous food resource. Forage was progressively depleted during the winter months after plant growth had ceased. Births were represented by transferring a fraction of maternal biomass to constitute the nucleus of the juvenile segment at the beginning of the growing season. Mortality losses depended on the extent of food depletion during winter, with population segments differing in their sensitivity to food deficiencies. This model replicated the sharply convex density dependence and consequent oscillatory dynamics shown by the sheep population, without incorporating weather variation. The demographic interpretation had ascribed this pattern to the high fecundity and rapid increase of the sheep population by as much as 50% during the course of summer, so that the abundance attained could greatly exceed what could be supported through the subsequent winter (Clutton-Brock and Coulson 2002). The metaphysiological model emphasized instead how the oscillatory dynamics were predisposed by the high forage quality on the wet and fertile island, with mortality precipitated by extreme depletion of this nutritious food at some stage of the winter. Contributing further was the absence of a buffer resource providing sufficient nutritional value to alleviate starvation.

For Soay sheep, incorporating demographic structure greatly improves the predictive success of models, largely because of the widely differing susceptibility of age/sex classes to starvation-related mortality (Coulson et al. 2001, 2008). For kudu, differences in the dynamics projected by an age-structured model and an equivalent unstructured logistic model appeared minor, because the rainfall influence dominated the dynamics (Fig. 7.7). The age-structured model indicated a more rapid rebound from population declines in dry years than the unstructured model, because the surviving animals consisted largely of prime-aged females with high reproductive potential.

Demographic assessments are generally focused on survival rates because this is the parameter included in matrix projection models. The metaphysiological approach represents instead the losses reducing biomass growth below some maximum potential through mortality, reductions in fertility, and increases in metabolic demands. The mortality response

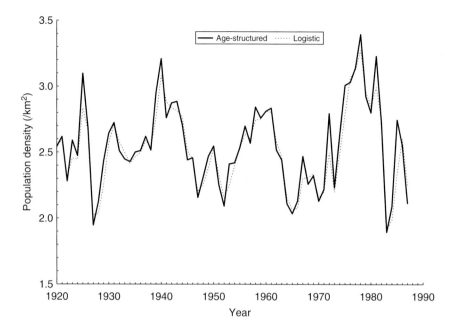

Figure 7.7 Projected dynamics of a kudu population in response to annual rainfall variation, comparing the output of an age-structured population model with that of the counterpart logistic model with K carrying capacity dependent on rainfall (from Owen-Smith 2007).

to rising density relative to resource production can appear effectively linear over a wide range, consistent with eqn. (7.8), allowing for some minimum mortality that must occur even at very low density (Figs. 7.8 and 7.9). For kudu, a curvilinear logistic function did not improve the fit to the data for either the adult female or juvenile segments. The truncation of annual mortality at an upper limit of 1.0 for elk in Yellowstone is an artifact of the time step spanned. Almost certain death within a month or two is a higher rate than certain death only over 12 months. Whether a linear or nonlinear relationship fits best depends on the threshold onset of mortality. For Soay sheep, a nonlinearly accelerating function seemed to provide a better fit to the mortality response for adult females, but not for males or juveniles.

Demographic segments, whether distinguished by age class (as in demographic models) or growth stage (as in metaphysiological models), are expected to differ in their mortality response to both rising density and diminishing food production, as postulated by Eberhardt (1977, 2002) and documented for kudu (Fig. 7.10). These segments differ also in their

(a)

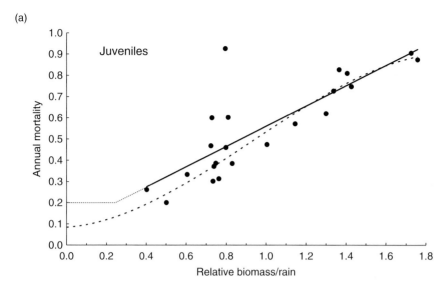

(b)

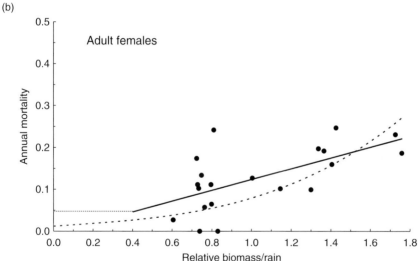

Figure 7.8 Mortality functions fitted to data for kudus, comparing fit of linear segments (*solid lines* plus hypothetical extrapolation) with that of logistic functions (*dashed lines*). (a) Juvenile mortality estimated from the mother–offspring ratio, assuming a minimum mortality loss of 0.2 due to predation or other causes at low density. (b) Adult female mortality encompassing both prime and old animals, assuming a minimum mortality loss of 0.05 per year due to animals reaching the maximum life span (from Owen-Smith 2000).

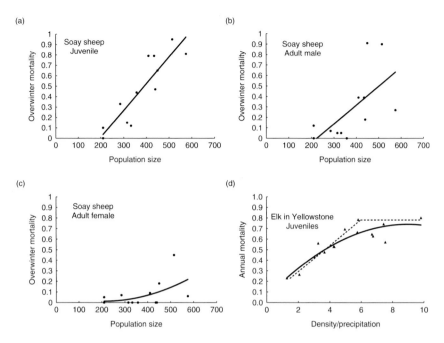

Figure 7.9 Stage-specific mortality responses to rising population density, controlled for weather influences. (a–c) Soay sheep on Hirta (from Milner et al. 1999); (d) Elk in Yellowstone (from Coughenour and Singer 1996).

vulnerability to predation. Knowing the age structure of the animals killed, not only the proportion of juveniles but also the proportion represented by older adults, is an important consideration in assessing the consequences for prey population dynamics (Vucetich et al. 2005). When mortality among prime females is elevated through predation, hunting, or management culling, fewer animals survive to old age, thereby reducing the effect of the senescent stage on population dynamics.

7.5   Adaptive responses and environmental contexts

The most detailed model resolution would be achieved by representing the responses of individual organisms, or classes of organisms with similar features, to their specific environmental and social contexts (Moen et al. 1997, 1998). However, such models have not yet made a major conceptual contribution. Adaptive responses are represented in a simpler way within metaphysiological models via shifts in resource exploitation among distinct habitats or regions at the level of aggregated population

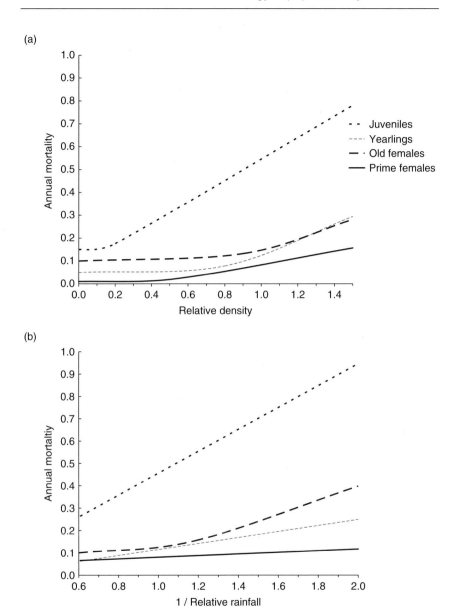

Figure 7.10 Comparative sensitivity of stage-specific mortality rates within a kudu population to rainfall variation relative to density variation (modified from Owen-Smith 2006).

biomass. This leads into an "ecology of equations," whereby the same basic model may generate different outcomes dependent on the environmental context, such as the nature and extent of the variability in resources and conditions (Owen-Smith 2002c). This could lead towards ecological niche modeling, assessing how resources and conditions influence where populations of specific species can persist (Soberon 2007), and further towards establishing the implications of climate and land-use changes for regional distribution patterns (Guisan and Thuiller 2005).

7.6   Summary and conclusions

Conventional modeling approaches accommodating only density feedbacks or the effects of demographic structure on population growth are inadequate to establish the consequences of environmental variability for population abundance. Weather conditions may impose additive mortality or influence the effective availability of food resources, while similar distinctions can be made with respect to predation. The metaphysiological modeling approach accommodates environmental variability through time and space by distinguishing processes reducing population growth below the maximum potential, as a result of within-year as well as between-year changes in resource fluxes and conditions. The specific influences of environmental conditions on particular age or stage classes can be incorporated as further refinements. Weather variation should be acknowledged as the music within this ecological theater, modulated by the spatial context, rather than being relegated to the status of noise.

Acknowledgments

Critical comments on various preceding drafts of this chapter were received from Tim Coulson, John Gross, Marco Festa-Bianchet, Tony Sinclair, John Fryxell, and Wayne Getz. None of them can be held responsible for the final shape that it has taken.

References

Andrew, M. H. 1988. Grazing impact in relation to livestock watering points. *Trends in Ecology and Evolution* 3: 336–339.
Arditi, R. and L. R. Ginzburg. 1989. Coupling in predator-prey dynamics: ratio dependence. *Journal of Theoretical Biology* 139: 311–326.

Berryman, A. A., J. Michalski, A. P. Gutierrez, and R. Arditi. 1995. Logistic theory of food web dynamics. *Ecology* 76: 336–343.

Bjornstad, O. N. and B. T. Grenfell. 2001. Noisy clockwork: time series analysis of population fluctuations in animals. *Science* 293: 638–643.

Brown, J. S., J. W. Laundre, and M. Guring. 1999. The ecology of fear: optimal foraging, game theory, and trophic interactions. *Journal of Mammalogy* 80: 385–399.

Caro, T. and P. Scholte. 2007. When protection falters. *African Journal of Ecology* 45: 233–235.

Caswell, H. 2001. *Matrix Population Models. Construction, Analysis, and Interpretation.* 2nd edition. Sinauer, Sunderland, Massachusetts.

Caughley, G. 1976. Plant-herbivore systems. In *Theoretical Ecology*, ed. R. M. May, pp. 94–113. Blackwell, Oxford.

Chamaille-Jammes, S., H. Fritz, M. Valeix, F. Murindagomo, and J. Clobert. 2008. Resource variability, aggregation and direct density dependence in an open context: the local regulation of an African elephant population. *Journal of Animal Ecology* 77: 135–144.

Clark, C. W. and M. Mangel. 2000. *Dynamic State Variable Models in Ecology. Methods and Applications.* Oxford University Press, Oxford.

Clutton-Brock, T. H. and T. Coulson. 2002. Comparative ungulate dynamics: the devil is in the details. *Philosophical Transactions of the Royal Society, Series B: Biological Sciences* 357: 1285–1298.

Coughenour, M. B. and F. J. Singer. 1996. Elk population processes in Yellowstone National Park under the policy of natural regulation. *Ecological Applications* 6: 573–593.

Coulson, T., S. Albon, J. Pilkington, and T. H. Clutton-Brock. 1999. Small-scale spatial dynamics in a fluctuating ungulate population. *Journal of Animal Ecology* 68: 658–671.

Coulson, T., E. A. Catchpole, S. D. Albon, et al. 2001. Age, sex, density, winter weather, and population crashes in Soay sheep. *Science* 292: 1528–1531.

Coulson, T., T. H. G. Ezard, F. Pelletier, et al. 2008. Estimating the functional form for the density dependence from life history data. *Ecology* 89: 1661–1674.

Coulson, T., J.-M. Gaillard, and M. Festa-Bianchet. 2005. Decomposing the variation in population growth into contributions from multiple demographic rates. *Journal of Animal Ecology* 74: 789–801.

Coulson, T., F. Guinness, J. Pemberton, and T. Clutton-Brock. 2004b. The demographic consequences of releasing a population of red deer from culling. *Ecology* 85: 411–422.

Coulson, T., P. Rohani, and M. Pascual. 2004a. Skeletons, noise and population growth: the end of an old debate? *Trends in Ecology and Evolution* 19: 359–364.

Dennis, B. and M. L. Taper. 1994. Density-dependence in time series observations of natural populations: estimation and testing. *Ecological Monographs* 64: 205–224.

De Roos, A. M., N. Galic, and H. Heesterbeek. 2009. How resource competition shapes individual life history for nonplastic growth: ungulates in seasonal food environments. *Ecology* 90: 945–960.

De Roos, A. M. and L. Persson. 2001. Physiologically structured models - from versatile technique to ecological theory. *Oikos* 94: 51–71.

East, R. 1984. Rainfall, soil nutrient status and biomass of large African savanna animals. *African Journal of Ecology* 22: 245–270.

Eberhardt, L. L. 1977. Optimal policies for conservation of large mammals with special reference to marine ecosystems. *Environmental Conservation* 4: 205–212.

Eberhardt, L. L. 2002. A paradigm for population analysis of long-lived vertebrates. *Ecology* 83: 2841–2854.

Festa-Bianchet, M. 1988. Seasonal range selection in bighorn sheep: conflicts between forage quality, forage quantity and predator avoidance. *Oecologia* 75: 580–586.

Festa-Bianchet, M., J.-M. Gaillard, and S. D. Cote. 2003. Variable age structure and apparent density dependence in survival of adult ungulates. *Journal of Animal Ecology* 72: 640–649.

Forchhammer, M., N. C. Stenseth, E. Post, and R. Langvatn. 1998. Population dynamics of Norwegian red deer: density dependence and climatic variation. *Proceedings of the Royal Society of London B: Biological Sciences* 265: 341–350.

Forsyth, D. M. and P. Caley. 2006. Testing the irruptive paradigm of large-herbivore dynamics. *Ecology* 87: 297–303.

Fowler, C. W. 1981. Density dependence as related to life history strategy. *Ecology* 62: 602–610.

Fritz, H. and P. Duncan. 1994. On the carrying capacity for large ungulates of African savanna ecosystems. *Proceedings of the Royal Society of London B: Biological Sciences* 256: 77–82.

Fryxell, J. M., J. F. Wilmshurst, and A. R. E. Sinclair. 2004. Predictive models of movement by Serengeti grazers. *Ecology* 85: 2429–2435.

Fryxell, J. M., J. F. Wilsmhurst, A. R. E. Sinclair, D. T. Haydon, R. D. Holt, and P. A. Abrams. 2005. Landscape scale, heterogeneity, and the viability of Serengeti grazers. *Ecology Letters* 8: 328–335.

Gaillard, J.-M., M. Festa-Bianchet, and N. G. Yoccoz. 1998. Population dynamics of large herbivores: variable recruitment with constant adult survival. *Trends in Ecology and Evolution* 13: 58–63.

Getz, W. M. 1993. Metaphysiological and evolutionary dynamics of populations exploiting constant and interactive resources: r-K selection revisited. *Evolutionary Ecology* 7: 287–305.

Getz, W. M. 1996. A hypothesis regarding the abruptness of density dependence and the growth rate of populations. *Ecology* 77: 2014–2026.

Getz, W. M. 1999. Population and evolutionary dynamics of consumer-resource systems. In *Advanced Ecological Theory*, ed. J. McGlade, pp. 194–231. Blackwell, Oxford.

Getz, W. M. and N. Owen-Smith. 1999. A metaphysiological population model of storage in variable environments. *Natural Resource Modeling* 12: 197–230.

Guisan, A. and W. Thuiller. 2005. Predicting species distribution: offering more than simple habitat models. *Ecology Letters* 8: 993–1009.

Hallett, T. B., T. Coulson, J. G. Pilkington, T. H. Clutton-Brock, J. M. Pemberton, and B. T. Grenfell. 2004. Why large-scale climate indices seem to predict ecological processes better than local weather. *Nature* 430: 71–74.

Hebblewhite, M., E. Merrill, and G. McDermid. 2008. A multiscale test of the forage maturation hypothesis in a partially migratory ungulate population. *Ecological Monographs* 78: 141–166.

Illius, A. W. and T. G. O'Connor. 1999. On the relevance of nonequilibrium concepts to arid and semi-arid grazing systems. *Ecological Applications* 9: 798–813.

Kernohan, B. J., R. A. Gitzen, and J. J. Millspaugh. 2001. Analysis of animal space use and movements. In *Radiotracking and Animal Populations*, eds. J. J. Millspaugh and J. M. Marzluff, pp. 126–166. Academic Press, San Diego , California.

Lande, R., S. Engen, B.-E. Saether, F. Filli, E. Matthysen, and H. Heimerskirch. 2002. Estimating density dependence from population time series using demographic theory and life-history data. *American Naturalist* 159: 321–338.

McNaughton, S. J. 1988. Mineral nutrition and spatial concentrations of African ungulates. *Nature* 334: 343–345.

Mech, L. D., R. E. McRoberts, R. O. Peterson, and R. E. Page. 1987. Relationship of deer and moose populations to previous winter's snow. *Journal of Animal Ecology* 56: 615–627.

Milner, J. M., D. A. Elston, and S. D. Albon. 1999. Estimating the contributions of population density and climatic fluctuations to interannual variation in survival of Soay sheep. *Journal of Animal Ecology* 68: 1235–1247.

Moen, R., Y. Cohen, and J. Pastor. 1998. Linking moose population and plant growth models with a moose energetics model. *Ecosystems* 1: 52–63.

Moen, R., J. Pastor, and Y. Cohen. 1997. A spatially explicit model of moose foraging and energetics. *Ecology* 78: 505–521.

Murray, N. G. and D. Brown. 1993. Niche separation of grazing ungulates in the Serengeti: an experimental test. *Journal of Animal Ecology* 62: 380–389.

Ogutu, J. O. and N. Owen-Smith. 2003. ENSO, rainfall and temperature influences on extreme population declines among African savanna ungulates. *Ecology Letters* 6: 412–419.

Owen-Smith, N. 1988. *Megaherbivores. The Influence of Very Large Body Size on Ecology*. Cambridge University Press, Cambridge.

Owen-Smith, N. 1990. Demography of a large herbivore, the greater kudu *Tragelaphus strepsiceros*, in relation to rainfall. *Journal of Animal Ecology* 59: 893–913.

Owen-Smith, N. 2000. Modeling the population dynamics of a subtropical ungulate in a variable environment: rain, cold and predators. *Natural Resource Modeling* 13: 57–87.

Owen-Smith, N. 2002a. *Adaptive Herbivore Ecology. From Resources to Populations in Variable Environments*. Cambridge University Press, Cambridge.

Owen-Smith, N. 2002b. A metaphysiological modelling approach to stability in herbivore-vegetation systems. *Ecological Modelling* 149: 153–178.

Owen-Smith, N. 2002c. Credible models for herbivore–vegetation systems: towards an ecology of equations. *South African Journal of Science* 98: 445–449.

Owen-Smith, N. 2004. Functional heterogeneity within landscapes and herbivore population dynamics. *Landscape Ecology* 19: 761–771.

Owen-Smith, N. 2005. Incorporating fundamental laws of biology and physics into population ecology: the metaphysiological approach. *Oikos* 111: 611–615.

Owen-Smith, N. 2006. Demographic determination of the shape of density dependence for three African ungulate populations. *Ecological Monographs* 76: 73–92.

Owen-Smith, N. 2007. *Introduction to Modeling in Wildlife and Resource Conservation*. Blackwell, Oxford.

Owen-Smith, N. and S. M. Cooper. 1989. Nutritional ecology of a browsing ruminant, the kudu, through the seasonal cycle. *Journal of Zoology* 219: 29–43.

Owen-Smith, N., G. I. H. Kerley, B. Page, R. Slotow, and R. J. Van Aarde. 2006. A scientific perspective on the management of elephants in the Kruger National Park and elsewhere. *South African Journal of Science* 102: 389–394.

Owen-Smith, N. and D. R. Mason. 2005. Comparative changes in adult versus juvenile survival affecting population trends of African ungulates. *Journal of Animal Ecology* 74: 762–773.

Owen-Smith, N. and M. G. L. Mills. 2006. Manifold interactive influences on the population dynamics of a multispecies ungulate assemblage. *Ecological Monographs* 76: 73–92.

Owen-Smith, N. and M. G. L. Mills. 2008. Shifting prey selection generates contrasting herbivore dynamics within a large-mammal predator-prey web. *Ecology* 89: 1120–1133.

Pulliam, H. R. 1988. Sources, sinks, and population regulation. *American Naturalist* 132: 652–661.

Royama, T. 1992. *Analytical Population Dynamics*. Chapman & Hall, London.

Rutherford, M. C. 1980. Annual plant production-precipitation relations in arid and semi-arid regions. *South African Journal of Science* 76: 53–56.

Scoones, I. 1995. Exploiting heterogeneity: habitat use by cattle in dryland Zimbabwe. *Journal of Arid Environments* 29: 221–237.

Smith, D. W., R. O. Peterson, and D. B. Houston. 2003. Yellowstone after wolves. *BioScience* 53: 330–340.

Soberon, J. 2007. Grinnellian and Eltonian niches and geographic distributions of species. *Ecology Letters* 10: 1115–1123.

Solow, A. R. 2001. Observation error and the detection of delayed density dependence. *Ecology* 82: 3263–3264.

Spalinger, D. E. and N. T. Hobbs. 1992. Mechanisms of foraging in mammalian herbivores: new models of functional response. *American Naturalist* 140: 325–348.

Stenseth, N. C. 1999. Predation cycles in voles and lemmings: density dependence in a stochastic world. *Oikos* 87: 427–461.

Stenseth, N. C., K.-S. Chan, G. Tavecchia, et al. 2004. Modelling non-additive and non-linear signals from climatic noise in ecological time series: Soay sheep as an example. *Proceedings of the Royal Society of London B: Biological Sciences* 271: 1985–1993.

Stenseth, N. C., M. O. Kittilsen, D. O. Hjermann, H. Viljugrein, and T. Saitoh. 2002. Interaction between seasonal density-dependent structures and length of the seasons explain the geographic structure of the dynamics of voles in Hokkaido: an example of seasonal forcing. *Proceedings of the Royal Society of London* B269: 1853–1863.

Stenseth, N. C., G. Ottersen, J. W. Hurrell, et al. 2003. Studying climate effects on ecology through the use of climate indices: the North Atlantic Oscillation, El Nino Southern Oscillation and beyond. *Proceedings of the Royal Society of London B: Biological Sciences* 270: 2087–2096.

Tremblay, J.-P., I. Thibault, C. Dusssault, J. Huot, and S. D. Cote. 2005. Long-term decline in white-tailed deer browse supply: can lichens and litterfall act as alternative food sources that preclude density-dependent feedbacks? *Canadian Journal of Zoology* 83: 1087–1096.

Turchin, P. 2003. *Complex Population Dynamics. A Theoretical/Empirical Synthesis*. Princeton University Press, Princeton , New Jersey.

Vucetich, J. A., D. W. Smith, D. R. Stahler. 2005. Influence of harvest, climate and wolf predation on Yellowstone elk, 1961–2004. *Oikos* 111: 259–270.

Walker, B. H., R. H. Emslie, N. Owen-Smith, and R. J. Scholes. 1987. To cull or not to cull: lessons from a southern African drought. *Journal of Applied Ecology* 24: 381–402.

Wang, G., N. T. Hobbs, R. B. Boone, et al. 2006. Spatial and temporal variability modify density dependence in populations of large herbivores. *Ecology* 87: 95–102.

Wilmers, C. C. and W. M. Getz. 2004. Simulating the effects of wolf-elk population dynamics on resource flow to scavengers. *Ecological Modelling* 177: 193–208.

Wilmshurst, J. F., J. M. Fryxell, B. P. Farm, A. R. E. Sinclair, and C. P. Henschel. 1999. Spatial distribution of Serengeti wildebeest in relation to resources. *Canadian Journal of Zoology* 77: 1223–1232.

Index

Note: page numbers in *italics* refer to figures, those in **bold** refer to tables